How We Can Save
Mayer Hillman
with Tina Fawcett

D0526992

Acknowledgements

We wish to pay a special tribute to Aubrey Meyer, the founder and director of the Global Commons Institute, for the inspiration of his ideas and the invaluable discussions held with him since 1990.

We also wish to record our particular appreciation of the guidance and editorial advice from Martin Toseland at Penguin, and the help and support of Heidi Hillman throughout the preparation of the manuscript. We are grateful to Saul Hillman for his contribution in developing the psychological themes in the book, especially for Chapter 4, and to Richard Blundel and Geraldine Fawcett for their comments on the drafts of the chapters.

Finally, the Polden Puckham Charitable Foundation is warmly thanked for its financial support and encouragement during the early stages of the research for the book.

Policy Studies Institute and University College London provided the research bases for the two authors. MH and TF

A note on information sources
Many of the facts and figures quoted in this book come from publications by the UK government and national and international organizations. A list of sources can be found in the chapter-by-chapter reference list at the end of the book.

Papers used by Penguin Books are natural, recyclable products made from wood grown in sustainable forests. The manufacturing processes conform to the environmental regulations of the country of origin.

Abbreviations

The following abbreviations and acronyms have been used throughout:

Units of measure
BTUs	British thermal units
GtC	gigatonne (thousand million tonnes) of carbon
GWh	gigawatt (thousand million watt) hours
kgC	kilogrammes of carbon
$kgCO_2$	kilogrammes of carbon dioxide
kWh	kilowatt hours
MtC	million tonnes of carbon
mtoe	million tonnes of oil equivalent

MW   megawatt (1 million watts)
ppm   parts per million
tC   tonnes of carbon
tCO_2 tonnes of carbon dioxide
toe   tonnes of oil equivalent

Note: Carbon dioxide emissions can be measured in either tonnes of carbon dioxide (tCO_2) or tonnes of carbon (tC). Tonnes of carbon is the most usual method of measurement in UK and international statistics and is used throughout the book, with the exception of Chapter 8, where emissions are measured in tonnes of carbon dioxide.

One tonne = 1,000 kg; one tonne of carbon = 3.67 tCO_2. To convert tC into tCO_2 figures, multiply by 3.67. To convert tCO_2 into tC, divide by 3.67. The same formula applies to kgC and $kgCO_2$.

Organizations
CfIT   Commission for Integrated Transport
DEFRA   Department for Environment, Food and Rural Affairs
DETR   Department for the Environment, Transport and the Regions
DfT   Department for Transport
DTI   Department for Trade and Industry
DTLR   Department of Transport, Local Government and the Regions
ENDS   Environmental Data Services
EU   European Union
GCI   Global Commons Institute
ICCEPT   Imperial College Centre for Energy Policy and Technology
IEA   International Energy Agency
IIED   International Institute for Environment and Development
IPCC   Intergovernmental Panel on Climate Change
IPPR   Institute for Public Policy Research
NGO   non-government organization
OECD   Organization for Economic Co-operation and Development
ONS   Office of National Statistics
OPEC   Organization of Petroleum Exporting Countries
PIU   Performance and Innovation Unit (Cabinet Office)
RCEP   Royal Commission on Environmental Pollution
SPRU   Science Policy Research Unit (University of Sussex)
UN   United Nations
UNEP   United Nations Environment Programme
WHO   World Health Organization

Other abbreviated terms

C&C	contraction and convergence
CFL	compact fluorescent lamps
CHP	combined heat and power
EEC	Energy Efficiency Commitment
GDP	gross domestic product
HEES	Home Energy Efficiency Scheme
IT	information technology
LPG	liquefied petroleum gas
NETA	New Electricity Trading Arrangements
PV	photovoltaics
VAT	value added tax

Part One: The Problem

1. Eyes Wide Shut
Introduction

Climate change is the most important issue of our age – perhaps of any age. If we, collectively and individually, do not act resolutely on it and to the extent that is imperative, the prospects are grim. Higher temperatures are predicted worldwide, with the *average* 6°C above current levels by the end of the century and higher still in some countries. Sea levels will rise inexorably and rainfall patterns will be greatly altered, with the frequency of drought conditions and severe flooding far more common. Large parts of the globe are likely to become uninhabitable, especially in delta and other low-lying regions of the world. We are already witnessing the first stages of a monumental catastrophe.

This is not the view of alarmists but the considered opinion of international climate scientists. It is already acknowledged by most governments around the world. The decisions that have to be taken in response to these growing threats must be made as a matter of urgency. The greenhouse gases we emit now, and have been emitting – mainly carbon dioxide from energy use – will remain in the atmosphere for hundreds of years, causing changes for thousands of years as the earth slowly reacts. Future generations will bear the heaviest burden for the present generation's irresponsibility. We are currently on the road to ecological Armageddon, with little apparent thought for the effects on ourselves, let alone on the populations succeeding us.

It does not have to be like this. Nor does anybody want it to be. As the UK government stated in 1990, it is 'mankind's duty to act prudently and conscientiously so that the planet is handed over to future generations in good order'. As well as posing the most demanding challenges to the character and quality of our way of life, therefore, the issue is foremost a moral one and has to be seen as such, however problematical that may be. Taking a moral position as a starting point – that it is a matter both of duty and necessity to try to save the planet – this book presents what we see as the only solution to climate change that has a realistic prospect of success. The direction is very simple and generally agreed: cuts must be made to greenhouse gas emissions. The difficult part, and where moral as well as scientific questions arise, is deciding by how much, by when and by whom. For instance, should it be the most 'energy profligate' nations and individuals who should be obliged to bear the greater burden in emissions reduction? We put forward a radical and innovative strategy that can achieve a sufficient decrease in emissions, by a set date, and in a way that is transparent and fair, and so can command wide public support. Adopting this strategy would mean the UK could both meet its international and national commitments to greenhouse gas reductions, and show global leadership.

We have written this book because the implications of climate change are not yet being taken seriously enough. People, including politicians and business leaders, talk about the importance of the environment, but greenhouse gas emissions around the world are accumulating in the atmosphere at an alarming rate. The apparent contradiction between belief and action may be because there is a feeling that climate change will turn out to be an ephemeral problem that will magically disappear or that human ingenuity will lead to the development of some miraculous technology so that our material standards of living can continue to rise for ever.

We challenge both these convenient myths. Evidence that climate change is already occurring, and will do so at a more rapid rate in the future, is increasing all the time. Technology, by itself, will not be sufficient to make the cuts required to prevent this change happening to an unimaginable extent. We do not discount the value of technological improvements but, far more importantly, we believe that radical change is required in people's behaviour. This will require individuals both to demand rigorous government intervention that will lead to carbon dioxide savings across society, and to make changes in their own lives to reduce their personal emissions of carbon dioxide.

A major focus in this book relates to the contribution that individuals in the

UK are making and could make. They are directly responsible for half of its emissions of carbon dioxide, the main greenhouse gas. Individuals must also take responsibility for the consequences of their actions and not look for excuses for evading it. We all have a crucial role to play. Deciding how much energy we use, in car travel, foreign flights or heating our homes, now has a moral dimension because of the carbon dioxide emissions they produce. This book provides the tools for individuals to audit their own emissions and gives information and advice on how to reduce them. Although it delivers some unwelcome messages, it is essentially optimistic. We believe that with the right information, motivation and emission-saving framework, individuals, the UK as a whole, and the world at large, will be able to avoid the extremes of climate change to which we are currently heading with eyes wide shut.

The unpalatable truths contained in the book may make uncomfortable reading. They challenge the deep complacency in our society that we can continue making energy-profligate lifestyle choices and 'get away with it', and that we can ignore the issue of the fair distribution, both internationally and between generations, of a commodity to which, in theory, everyone has an equal claim.

There is one simple fact underlying all of this: we now know that the planet has a finite capacity to absorb greenhouse gas emissions. This means that there *are* limits to the growth of energy-dependent activities if excessive stress on the ecology of the planet and on its future populations is to be avoided. We must come to terms with the inherent contradictions of pursuing a policy of economic growth (based on increasing energy use) and, at the same time, of preventing these serious consequences of climate change. We cannot merely rely on optimism: action must be taken by everyone *now*.

By the time you have finished reading this book, the following twelve key points will have been covered:

1. Why the threat posed by climate change to human welfare and the environment, both in the UK and worldwide, is so grave and immediate.
2. How our use of fossil-fuel energy is the main source of the threat.
3. What we use energy for, and the forces that are driving its consumption ever upwards.
4. What excuses people use to avoid taking climate change seriously and why these lack validity.
5. Why our current collective response to the threat of climate change and its implications is totally inadequate.
6. Why the technological options for reducing carbon dioxide emissions, such as greater energy efficiency and far more use of renewable energy, are limited in scope.

7. Why the principle of equity must be applied in international negotiations on reducing greenhouse gas emissions.

8. How a system of carbon rationing for individuals based on this principle, and carbon caps for business and the public sector, would ensure that each country contributes its fair share in a global agreement.

9. How a relatively painless transition towards the necessary target can be achieved.

10. What the average UK personal ration must be, how it can be reduced to that level, and what we can do as individuals to audit and reduce our own emissions.

11. Who the winners and losers would be under the system of carbon rationing.

12. Why complacency and procrastination on the issue of climate change must stop.

Hopefully you will agree with the line of argument set down in the book and will be encouraged to join in promoting a radical reappraisal of personal and public decisions from a climate change perspective. Individuals need not only to adapt their lifestyles but, more importantly, to press for the political change that is the only way out of the impasse into which our head-burying instincts have led us. Widespread public support is vital now. Time is running out!

2. Beyond the Planet's Limits
Climate Change: Why, How and What Next?

Climate change is the most serious environmental threat that the world has ever faced. The dangers can hardly be exaggerated: within one hundred years, temperatures could rise by 6°C worldwide, much of the earth's surface could become uninhabitable, and most species on the planet could be wiped out. In the UK, during the next fifty years, we will increasingly experience more heatwaves, higher summer temperatures, fewer cold winters, drier summers, wetter winters and rising sea levels. As a consequence, millions of people will be at high risk from flooding, there will be thousands of deaths from excessive summer temperatures, diseases from warmer regions will become established, some species and habitats will be lost for ever and patterns of agriculture and business will have to change radically.

Why is the climate changing?

The climate is changing because the natural mechanism known as the 'green-house effect' which acts to warm the earth is being increased by human-induced emissions of greenhouse gases. As the concentrations of the emissions rise well above their natural levels, additional warming is taking place.

To explain this effect in a little more detail, the temperature of the earth is determined by the balance between energy coming in from the sun in the form of sunlight (visible radiation) and energy constantly being emitted from the earth into space. The energy coming in from the sun can pass through the atmosphere almost unchanged and warm the earth, but the heat (infrared radiation) emanating from the earth's surface is partly absorbed by certain gases in the atmosphere and some of it is returned to earth. This further warms the earth's surface and the lower strata of the atmosphere. Without this natural greenhouse effect, the planet would be over 30°C cooler than it is now and would be too cold for us to inhabit. However, the greenhouse gases we add to the atmosphere mean that more heat is being trapped. This is leading to global warming (higher global temperatures) and other changes to the climate.

The primary cause is our use of fossil-fuel energy (coal, oil and gas). This is because burning fossil fuels, which are carbon-based, results in the production of carbon dioxide. Globally, carbon dioxide contributes to more than two-thirds of the warming; and in the UK, it accounts for five-sixths of emissions of greenhouse gases. Due to their chemical structure, different types of fuel give rise to different amounts of carbon dioxide per tonne burned and per unit of energy produced. Coal is the fossil fuel which then produces the most carbon dioxide per unit of energy produced, followed by oil and gas. (Energy use is explored in detail in Chapter 3.)

In addition to fossil-fuel combustion, land use changes also contribute to the release of carbon dioxide into the atmosphere. These changes include clearing land for logging, ranching and agriculture, or switching from agricultural to industrial or urban use. Vegetation contains carbon that is released as carbon dioxide when it decays or burns. Normally, lost vegetation would be replaced by regrowth with little or no extra emissions because the replacement vegetation absorbs carbon dioxide from the atmosphere as it grows. However, over the past several hundred years, deforestation and other land use changes around the world have contributed to one-fifth of the additional carbon dioxide in the atmosphere that is attributable to human activity, mostly through cutting down tropical forest.

In addition to carbon dioxide, there are five other important greenhouse gases: methane, nitrous oxide, hydrofluorocarbons, perfluorocarbons and sulphur hexafluoride. The most significant of these are the first two. Methane emissions come primarily from agriculture, waste, coal mining and natural gas distribution. They can be a major component of greenhouse gas emissions in countries with strong agricultural economies. For example, as a by-product of their digestion, New Zealand's 45 million sheep and 8 million cattle produce about 90 per cent of the country's methane emissions, which equates to over 40 per cent of the country's total production of greenhouse gases. Nitrous oxide is generated from agriculture, industrial processes and fuel combustion. The other greenhouse gases are emitted from a small range of industrial processes and products. With the exception of methane, these other gases are much easier to control through technological change than is carbon dioxide.

This book concentrates on carbon dioxide emissions from fossil-fuel use because this is the largest global source of greenhouse gas emissions.

Carbon dioxide emissions

Concentrations of carbon dioxide in the atmosphere are increasing, and have been doing so since the Industrial Revolution. They have increased from 280 parts per million (ppm) in 1750 to 373 ppm in 2002, a rise of a third. The staggering increase since 1959 as measured at Mauna Loa in Hawaii (the meteorological station with the longest continuous recording of atmospheric carbon dioxide concentrations available in the world) is shown in Figure 1 (see below). In 1997–8, there was an increase of 2.87 ppm, the largest single yearly jump ever recorded there.

Earlier data have been obtained from measurements of air that has been trapped in ice over thousands of years. These data reveal that today's carbon dioxide concentration has not been exceeded in the past 420,000 years and probably not during the past 20 million years. The rate of increase over the past century is also unprecedented. Compared to the relatively stable carbon dioxide concentrations (about 280 ppm) of the preceding several thousand years, the increase during the industrial era is disturbing.

The carbon dioxide emissions leading to these concentrations can be seen in Figure 2 (see below). They show dramatic and accelerating growth, as the carbon deposits laid down over millions of years as fossil fuels are released into the atmosphere. Half the total emissions since 1750 have occurred since the mid-1970s. Annual emissions have doubled since the mid-1960s and trebled since the mid-1950s – a tripling in less than fifty years. Future scenarios suggest that annual emissions of carbon dioxide could be up to five times their current level by 2100, resulting in up to 6°C of global warming.

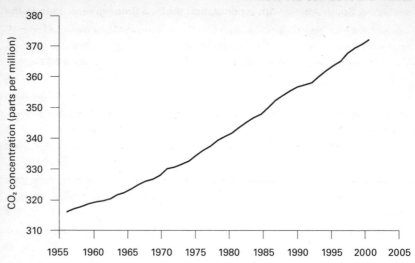

Figure 1: Atmospheric carbon dioxide (CO_2) concentrations since 1959, Mauna Loa

At the end of the twentieth century, total global emissions from fossil fuels amounted to about 6,600 million tonnes of carbon (MtC). Emissions from North America (USA, Canada and Mexico) made up a quarter of the total, and those from Western Europe accounted for about one-seventh.

The UK's carbon dioxide emissions in 2002 were 150 MtC – over 2 per cent of the world total. They exceed those produced by the whole African continent,

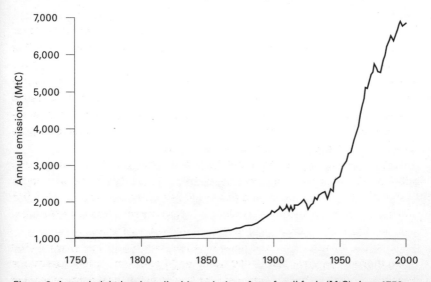

Figure 2: Annual global carbon dioxide emissions from fossil fuels (MtC) since 1750

excluding South Africa, with a population thirteen times greater than in the UK. However, contrary to the global trend, UK emissions have actually been falling over recent years: in 2002, they were 5 per cent lower than in 1990, and lower by about 20 per cent than those in 1970. As is more fully explained in Chapter 3, this is not because our energy consumption has decreased. The main reason is because the fuels used for electricity generation and heating have changed to ones which emit less carbon dioxide per unit of energy.

On average, about 1.1 tonnes of carbon are emitted for each person in the world at present. If carbon dioxide emissions are compared by country, the differences are stark. Near the top of the league is the United States, at 5.5 tC per person (some oil-producing nations have higher levels). In the UK, they are about 2.5 tC on average, which is around two and a half times the global average. The developing nations currently contribute much less – with China's emissions at 0.6 tC per person and India's at 0.3 tC. Afghanistan is at the bottom of the emissions league, at 0.01 tC, just one-hundredth of the global average and less than one five-hundredth of people in the USA.

Partly because the UK was the first country to industrialize its economy and use fossil-fuels, our contribution to the total emissions since 1750 is much higher than our current contribution: overall, it has been responsible for 15 per cent of the global figure. By comparison, the USA has been responsible for 29 per cent of global emissions over the same period.

How is the climate changing?
Temperature changes

Figure 3 (see below) shows how global-average surface temperature (over sea and land) has risen from 1850 to the present day. The data are set out in terms of the 'anomaly', that is the difference between each year and the average temperature in the period 1961–90. Before 1978, it was generally colder than the 1961–90 average, with all later years being warmer. Because temperatures vary naturally from year to year, climate scientists must compare several years' temperature records with long-term averages to be sure the suspected temperature change is significant.

Global temperature rose by about 0.6°C during the twentieth century, with about 0.4°C of this warming occurring since the 1970s. The global average is based on millions of individual measurements taken from around the world and this temperature record is considered by climate scientists to be the most reliable information describing the state of the global climate.

The 1990s was the warmest decade of the twentieth century, with 1998 being the warmest year since 1860 when world temperature measurements began. Scientists at the University of East Anglia, in the most comprehensive study to date of climatic history, have confirmed that the earth is now warmer than it has been at any time in the past 2,000 years. And there is no sign of this trend reversing: 2002 proved to be the second warmest year on record. Temperatures on land have warmed more than the oceans: they rose in central England by almost 1°C during the twentieth century, and the 1990s was the warmest decade in that part of the UK since measurements were first taken in the 1660s. On 10 August 2003, a new UK record temperature of 38.5°C was recorded in Kent, exceeding by 1.4°C the previous record set in 1990.

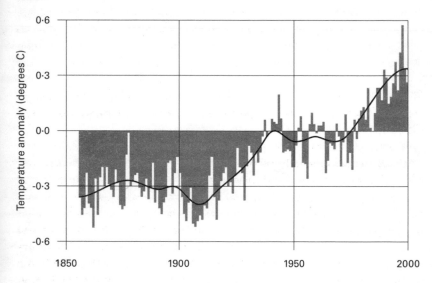

Figure 3: Changes in global-average surface air temperature compared with the long-term average for 1961–90

Other climate changes

The climate system is driven by energy from the sun. Rising temperatures increase the amount of energy in the system and this has knock-on effects on many other aspects of climate. As a result of increasing temperatures, climate models predict changes in rainfall amounts and patterns, and increased occurrence of storms, heatwaves and other extreme events. Some of these expected changes can already be seen in the UK and around the world. The well-known variability of the climate can make it difficult to be confident that individual events are unusual and are specifically caused by climate change. However, as Mike Hulme, a leading British climate scientist, has stated: 'There is no longer such a thing as a purely natural weather event.'

In the UK, there is strong evidence of changing rainfall patterns and new climatic extremes:

- A larger proportion of winter precipitation in all regions now falls on isolated days of heavy rainfall than was the case fifty years ago.
- Winters over the last 200 years have become wetter relative to summers.
- Heatwaves have become more frequent in summer.
- There are now fewer frosts and cold spells in winter.

These changes have both immediate and indirect effects on people and the natural environment. Changing winter rainfall patterns, for example, increase the risk of flooding and this has been exceptional in the UK and across Europe in recent years. The floods of Easter 1998 that struck in the Midlands, the Thames Valley, East Anglia and Wales were the worst for 150 years, with thousands of families forced out of their homes. Although nobody can be certain that global climate change caused these floods, they are the sort of events that would be expected under climate change.

Evidence for other changes in climate is not so strong – yet. There is particular concern about the potential increased frequency of extreme events, such as storms, as these tend to have severe effects on the population and the built environment as well as, of course, the natural environment. A further possibility is that global climate change may be having an effect on other weather systems. There is suspicion, though not yet proof, that the strongest El Niño event (a complex natural change in weather patterns which recurs every few years, affecting the equatorial Pacific region and beyond) in the twentieth century, which caused worldwide damage valued at $32 billion in 1997/8, was made worse by global warming.

Effects of climate change on the natural and human environment
Glaciers and sea ice

Climate change is now wide-ranging in its impacts. Higher temperatures have already had a measurable effect on land glaciers and sea ice. Mountain glaciers have been shrinking in almost all areas of the world. For example, the glacier from which Edmund Hillary and Tenzing Norgay began their ascent of Everest in 1953 has retreated by about three miles over the past fifty years. There has been a substantial thinning of Arctic sea ice in late summer: in August 2000, there was no ice at the North Pole – it was in a stretch of open water. More generally, sea ice in the Northern Hemisphere has decreased by an area the size of the UK (quarter of a million square kilometres) per decade since 1972.

Sea levels and wave heights

Other effects on the physical environment include rising sea levels, due to both the expansion of warmer water in the oceans and the additional water produced by the melting land glaciers. After adjusting for natural land movements, the average sea level around the UK is now about 10 cm higher than it was in 1900, and Atlantic waves hitting the western and southern coasts of the British Isles are over 1 metre higher. Increasing wave heights are accelerating coastal erosion, and rising sea levels are making coastal areas more vulnerable to flooding. As a consequence, the Thames Barrier, completed in 1982 to protect London from any foreseen flooding in the future, is already considered inadequate. Normally it is closed just two or three times a year, but from November 2000 to March 2001 it was closed twenty-three times.

Flora and fauna

Global climate change is also leading to changes in the seasonal behaviour and geographical location of fauna and flora. The change in temperature is equivalent to the UK moving south at 20 metres a day. The temperature rise of 1°C since 1900 in central England has resulted in a longer growing season for plants. Plants now grow for an extra month every year, with spring starting earlier and winter coming later. The records of an Oxfordshire naturalist also reflect this: the 385 wildflower species that were monitored bloomed on average four and a half days earlier in the 1990s compared with forty years previously, with one species now flowering fifty-five days earlier. Another recent striking illustration of this change was that, in 2002, spring was judged to have arrived three weeks early. Butterflies, birds and fish are already moving to new habitats to survive. Since 1980, eighteen new fish species from southerly waters have been caught in the now warmer sea off the Cornish coast. Also, there is evidence of a decrease of species attuned to northerly conditions: a combination of over-fishing and global warming has led to cod stocks in the North Sea falling to one-tenth of what they were thirty years ago. The North Sea has warmed by 3°C and, as the cod is a cold-water fish, the population left after over-fishing appears to be migrating northwards. With normal seasonal patterns being disrupted as the climate changes, the UK is gaining some species and is likely to lose others.

What might happen next?

What happens next with climate change depends on the following:

- The developing reaction of global systems to the gases already emitted.
- How much more carbon dioxide and other greenhouse gases are released into the atmosphere.
- How the climate reacts to additional greenhouse gas emissions, and, in particular, whether there are any 'unexpected' events or effects in which higher temperatures set in train a process that raises the temperature still further, in an upward spiral.

The financial costs alone are likely to be enormous. The economic damage caused by flooding worldwide in the fifty years since 1950 is estimated at almost $300 billion.

We can't just switch off climate change

Climate change does not have an instantaneous 'off-switch'. Climate change and its effects cannot quickly be reversed by reducing or even eliminating future emissions of greenhouse gases. This is for two reasons. First, greenhouse gases released into the atmosphere don't just disappear straight away. They linger for decades (in the case of relatively short-lived gases like methane), or hundreds of years (for carbon dioxide), or even thousands of years (for the long-lived gases like perfluorocarbons). Even if no additional carbon dioxide were emitted from now on, atmospheric concentrations would take centuries to decline to pre-Industrial Revolution levels.

Second, the reason that the effects of climate change cannot be simply reversed is that the planet reacts to changes in temperature over a long timescale. Delayed effects from the warming caused by the emissions already in the atmosphere will stretch way into the future. Global increases in mean surface temperature, rising sea levels from thermal expansion of the ocean, and melting ice sheets are projected to continue for hundreds of years. Even if all emissions of greenhouse gases ceased tomorrow, the climate would continue to change and, with it, its effects on life on the planet. We are changing the climate not just for generations, but for tens of generations to come.

The persistence of greenhouse gases in the atmosphere should determine the action we need to take. Stabilization of carbon dioxide emissions at near-current levels would not lead to the stabilization of atmospheric concentrations. These would go on rising – the 'old' carbon dioxide being joined by the 'new'. Stabilizing concentrations – at any level – requires the eventual reduction of global emissions to a fraction of their current levels. How small that fraction should be remains uncertain. However, over the next hundred years or so, they will need to decline to the level that does not exceed the capacity of natural land and ocean sinks to absorb them. This amount is in the region of 200 MtC per annum – that is at about the UK level rather than the current *global* emissions of carbon dioxide!

No shortage of fossil fuels

There is a popular misconception that we need to worry about the world running out of fossil fuels. But long before fossil fuels run out, the effect of continuing to use them at current rates will cause havoc to the climate and the planet, and that should be the greater cause for concern. Total carbon dioxide emissions from fossil-fuel use since 1751 are estimated at 280 GtC (thousand million tonnes of carbon). Total reserves on earth (including those not yet discovered) are estimated to amount to 5,000 GtC. Therefore stocks of fossil fuel have the potential to emit about eighteen times more carbon dioxide than has been emitted over the past 250 years. Yet governments around the world continue to act as if the exhaustion of fossil-fuel reserves is a serious problem when precisely the opposite is true. Using even a fraction of what has already been discovered would spell the end of a habitable planet for most of its population.

Sheikh Yamani, a founding architect of the Organization of Petroleum Exporting Countries (OPEC), has reportedly said: 'The Stone Age came to an end not because we had a lack of stones, and the oil age will come to an end not because we have a lack of oil.' It is essential that the significance of this observation is properly acknowledged if we are to prevent disastrous climate change.

Future possibilities

The Intergovernmental Panel on Climate Change (IPCC) is an international body of scientists and other experts which was formed to provide information and advice on climate change, and its reports have become the standard works of reference on the subject. It acted as the scientific advisory body for the Kyoto Protocol (see below). To consider the range of future increases in energy use and subsequent carbon dioxide emissions, it has developed four 'families' of scenarios. These scenarios portray the world with different combinations of demographic change, social and economic development, and broad technological developments. All the scenarios assume no action is taken to combat climate change. The results are used to predict how much global warming could occur in the future.

The extent to which energy use increases and to which carbon dioxide emissions and global warming rise depends on which scenario is being analysed. Energy demand during this century grows in all of the four scenarios, from one and a half to six times that in 1990. The differences in energy demand and the types of energy used translate into a range of carbon dioxide emissions, in turn leading to different concentrations in the atmosphere. By 2100, these vary between 500 and 950 ppm, depending on the scenario – remember that pre-industrial levels were just 280 ppm. The greater the concentration of carbon dioxide, the greater the degree of global warming. Without action to reduce the emissions, the IPCC does not foresee a scenario in which concentrations can be kept below 500 ppm by 2100. Although it does not attach probabilities to any of these scenarios, the scenario with most in common with a 'business as usual' development of current trends results in the highest concentrations of carbon dioxide – 950 ppm. In other words, the world's current development path is the one likely to cause most damage to the climate.

From regrettable to tragic: predictions of future effects of climate change
Higher temperatures

The IPCC's projection, published in 2001, is that the world's temperature is likely to increase by 1.4–5.8°C over the period 1990 to 2100, depending on which greenhouse gas emission forecast is closest to the truth. However, as this is a projection for the *average* global surface temperature (which includes sea surface), it is very likely that nearly all land areas will warm to a greater extent. This is particularly true in northern North America and northern and central Asia where land temperatures could rise by 10°C. The latest reports show that the effects of greenhouse gas emissions on temperature could be even more extreme than this. A new climate-modelling approach, developed at the UK's Hadley Centre for Climate Prediction and Research, suggests that the twenty-first century could see more and faster warming than that estimated by the IPCC. It is important to put possible temperature changes in perspective. The global temperature difference between the last Ice Age and the present time was around 5–6°C. An increase of 10°C would create conditions in the south of England comparable to the Sahara Desert. Warming by 5–10°C is therefore a terrifying prospect.

Global impacts

The likely effects on humans and the natural environment of the scenarios based on high emissions range from the death of coral reefs to the creation of millions of environmental refugees. The New Economics Foundation suggests as many as 150 million people may be forced from their homes by climate change by 2050. Many species will become extinct: it is expected that by 2060 Arctic pack ice will have melted so much that all polar bears will starve because the animals they feed on, such as seals, will become scarce. A huge wave of extinction could be set in motion before the end of the century: researchers at Bristol University have shown that 6°C of global warming was enough to wipe out up to 95 per cent of the species alive on earth at the end of the Permian period, 250 million years ago. Infectious diseases will rise as the world gets warmer and, in addition, they will sweep north into higher latitudes.

Humans may not become extinct, but far more of us will die prematurely and we are very likely to run out of comfortable areas of the world in which to live. The Earth Policy Institute in Washington estimated that 35,000 people died in Europe during the record heatwave in the summer of 2003 (including 2,000 in the UK). And a World Health Organization (WHO) report estimated that over 150,000 people in developing countries are now dying each year from the effects of global warming, ranging from malaria, malnutrition, extremes of heat and cold, to floods and that by 2020 this number will have almost doubled.

Many countries will be under threat from rising sea levels, drought, storms, heatwaves and extreme economic and social disruption. Sea levels are predicted to rise by a metre over the next century, leading to heavily populated delta areas of the world such as in Bangladesh and China becoming submerged. The likelihood is that the frequency of heatwaves in the UK, like the record one in 2003, will increase.

The consequences of allowing the higher emission scenarios to become reality are global, irreversible and catastrophic. How can we allow this to happen given that we know what needs to be done? Our lack of action or sense of urgency is criminal and immoral and will, without doubt, be justifiably condemned by future generations.

Impacts on the UK

The climate in the UK is expected to be very different towards the end of the twenty-first century under all of the IPCC emission scenarios. The changes may not be as severe as in some other areas of the world. However, they will be sufficient to alter our way of life completely and to require alteration to many aspects of the economy. The climate will be warmer, and wetter in winter and drier in summer. By the 2080s, a daytime summer temperature might be expected to exceed 42°C once a decade in lowland England. The present record temperature for the UK is 38.5°C. Snowfall amounts will decrease, and large areas of the UK are likely to experience long sequences of winters without snow, with the Scottish skiing industry an obvious early casualty.

Other predicted consequences range from the regrettable – grass lawns no longer being viable in southern England and centuries-long patterns of gardening having to change – to the tragic, with outbreaks of malaria and thousands of deaths per year from heatstroke (though the warmer winters will reduce deaths from cold temperatures). The Foresight Flood and Coastal Defences Project predicts that by 2050 more than 3.5 million people in the UK will be 'at high risk' of flooding in their homes. In addition, two-thirds of the coastline of England and Wales will be subject to increased erosion owing to a combination of rising sea levels, sea surges and changes in the shape of the coastline. The annual economic damage from flooding will rise from £1 billion to up to £20 billion, depending on how much is spent on flood defences over the next few decades. Ultimately climate change will affect agriculture, water resources, human health, wildlife and the countryside, and will be highly expensive to deal with. But this is as nothing compared to the global effects of climate change.

The uninhabitable planet?

All the predictions above are based on the climate continuing to respond in a predictable way to increasing temperatures, with no unexpected shocks or 'positive feedback' (self-reinforcing effects which accelerate climate change) in the climate system. However, current understanding of the climate system is not reliable enough to be sure that something unexpected will not take place. In fact, geological records show that fast changes in climate have occurred in the past. Given this, several even more worrying scenarios have been advanced, including the following:

- The benefits of the Gulf Stream, which transports warm water past the UK and which provides its relatively mild winters, could be reduced or even lost. This could lead to Western Europe cooling by 5°C within a few decades – a return to the Ice Age.
- Vast quantities of methane (a greenhouse gas twenty times more potent than carbon dioxide) are stored in sediments below the shallow seabed of the Arctic, according to the US Geological Survey, as well as below the

tundra in northern Canada and Siberia. If the temperature surrounding the methane warms, it becomes unstable and methane gas is released, causing temperatures to increase further. This would create a positive feedback with a very real risk of global warming developing totally beyond control.

Many other unexpected events that we would be unable to control could happen. As the UK's then environment minister, Michael Meacher, wrote in early 2003: 'We don't know the limits of nature – how much rain could fall for how long a period, how much more powerful and frequent hurricanes could become, for how long droughts could endure. The ultimate concern is that if runaway global warming occurred, temperatures could spiral out of control and make our planet uninhabitable.'

The world's response
The Kyoto Protocol

The world's governments have responded to the threat of climate change. In 1992, the United Nations Framework Convention on Climate Change was created. Its objective is for the world to achieve stabilization of 'greenhouse gas concentrations in the atmosphere at a level that would prevent dangerous human-induced interference with the climate system'. However, the convention defined neither what level of carbon dioxide concentrations in the atmosphere would be dangerous nor did it set an upper limit.

In order to put the objective of the UN convention into action, the Kyoto Protocol was created. It was designed to be the first legally binding treaty aimed at cutting emissions of the main greenhouse gases. More than 150 nations signed it in 1997. However, the protocol has not yet been 'ratified'; that is, it is not yet legally binding. The USA, the world's biggest carbon emitter, declared in 2001 that it was no longer prepared to sign. In September 2003, Russia again delayed ratifying the treaty; without Russian agreement, the treaty will not have sufficient signatories to go ahead. Under its terms, industrialized nations ('Annex 1' countries, in the jargon) committed themselves to a range of targets to reduce emissions during the twenty-year period between 1990, the base year, and 2010. (Strictly speaking, the reduction target applies to average emissions of the five years 2008–12, but this is simplified here to 2010.) At present, non-industrialized countries do not have targets.

As part of the Kyoto Protocol, the member states of the EU jointly agreed to undertake an 8 per cent reduction of the six key greenhouse gases by 2010. The intention is to achieve this by a mixture of higher or lower reductions for some nations and a set of maximum increases for others. The UK target is a reduction of 12.5 per cent under the EU burden-sharing arrangements.

The net effect of the treaty, if the targets are met, will be to reduce industrialized countries' emissions by 5 per cent. The protocol's scientific advisers, the IPCC, say this will delay the effects of climate change by, at most, ten years. Clearly the importance of the Kyoto Protocol, such as it is, is more symbolic than realistic. The hope has to be that the agreement is the first step in a series of future international treaties which will bring about significant reductions in greenhouse gas emissions to limit damage from climate change and achieve the ultimate goal of climate stabilization. In theory, by operating under the protocol, countries will learn how to reduce emissions at least cost, and will set up emissions trading systems to help them to do this. If the rules are designed correctly, emissions trading should mean that countries with cheaper emissions-saving options will do more than their agreed share, and will then sell their 'spare' emissions reductions to countries where national reductions are more expensive. In addition, industrialized countries will transfer funds and technologies to developing countries to help them find cost-effective ways of achieving lower emissions. If it works as planned, the Kyoto Protocol will make a transition to a 'low carbon' world easier.

Unfortunately, it is unlikely that the protocol will achieve its aims, let alone lead the way to future treaties. The decision of the USA not to ratify the Kyoto treaty was a major blow to the process. Nor is there much sign that the USA is taking unilateral domestic action to reduce its emissions in line with what it might have agreed to under the treaty. Not having the world's biggest polluter on board undermines the authority and effectiveness of the protocol.

It is a cause for concern also that many of the countries that have agreed to the Kyoto Protocol are not on course to achieve their targets. In the EU, the latest assessment is that ten of the fifteen member states are likely to miss them by a wide margin. Even those countries that were allowed significant increases in emissions under the EU burden-sharing arrangements, including Portugal, Spain and the Republic of Ireland, are failing in their commitments. There is still just about time for the EU to meet its targets if strong action is taken, but at present this does not seem to be a political priority. The Kyoto treaty looks set to fail.

The underlying reason for pushing the agreement through was supposed to be to lay the foundation for future treaties and more ambitious reductions, rather than to achieve the fairly minor reductions in greenhouse gas emissions on which it compromised. Instead, the biggest carbon dioxide producer in the world has not joined the treaty, the treaty has not yet been ratified six years after being signed, and the majority of supposedly supportive countries are highly likely to miss their targets by a wide margin. Rather than being a symbol of the determination of the world to tackle climate change, it now appears as a symbol of precisely the opposite, of the short-term (perceived) economic interests of a few countries taking priority over the long-term future of the whole world.

In addition to its obligations under the Kyoto Protocol, the UK has set further relatively ambitious targets. On the basis of a feasibility analysis carried out by the Royal Commission on Environmental Pollution (RCEP), the current Labour government has committed itself to reduce carbon dioxide emissions by 20 per cent by 2010 from their level in 1990. It has also declared that it is aiming for a 60 per cent reduction by 2050, achievable in the view of the IPCC by using existing technology but nevertheless representing an even more important advance on the Kyoto treaty. In setting this target, the UK government has shown admirable global leadership. However, announcing a target is only the start of a long transitional process towards fulfilment. (The difference between this ambition for the future and current, often contrary, trends is explored in succeeding chapters.)

Is there a 'safe' limit for carbon dioxide emissions?

The international community has agreed that concentrations of greenhouse gas emissions in the atmosphere should not be allowed to reach 'dangerous' levels, but has not said what these might be. A limit of 550 ppm for carbon dioxide emissions has been suggested by several bodies, including the EU Council of Environment Ministers. However, it has not been universally accepted: 550 ppm is twice the level that was in the atmosphere prior to the Industrial Revolution, and current understanding of the way in which the climate and natural systems work may not be reliable enough to guarantee that the degree of change under these conditions would be safe and acceptable.

Members of the RCEP also recommended 550 ppm as an upper limit and this too has been accepted by the UK government. However, taking this figure as the maximum that can be contemplated presupposes that uncontrollable positive feedback, referred to earlier, is not set in train before this level of concentration is reached. A lower limit of 450 ppm would be a more risk-averse maximum. To stay under a cap of 450 ppm would require UK emissions to reduce by 80 per cent from 1997 levels by 2050 and 90 per cent by 2100, as its part of a global agreement. But, although it seems a preferable target, even 450 ppm may turn out not to be a truly safe limit. The target may have to be ratcheted downwards as more evidence of the process of climate change emerges. Of course, the only truly safe limit would be a return to pre-industrial levels of emissions, but this is unlikely ever to be achieved. (These issues are discussed in more detail in Chapter 7.)

The Global Commons Institute (GCI), a very early and continuing contributor to proposing means of resolving the problems of climate change, has argued for many years that success can only be achieved as part of a global agreement to restrict emissions worldwide, and one in which everyone in the world is entitled to emit equal amounts of greenhouse gases. Under its framework proposal, emissions from developed countries would decrease most, while those from some developing countries would be allowed to rise.

Sceptics

There are critics of both climate change science and the actions being taken or contemplated by the international community in order to try to reduce greenhouse gas emissions. However, most prominent critics are not climate experts. Sir John Houghton, previously co-chairman of the scientific working group of the IPCC, has estimated that there are fewer than ten active research scientists who disagree with the notion of human-induced climate change. This compares with the hundreds of climate scientists and thousands of other experts contributing to the IPCC research. In fact, the argument is really about response to risk, not about science. Sceptics suggest that because the global climate system is imperfectly understood (something climate scientists do not dispute), it is too soon to be sure that changes in global temperature are being interpreted correctly. Moreover, they argue that current models do not accurately represent the climate, adding still further uncertainty and thereby providing a case for not acting as a matter of urgency.

However, the world community has clearly decided that the science on this subject is convincing enough to warrant an international treaty (the Kyoto Protocol) to begin to address the problem. Sceptics may have spread doubt and delayed serious action to mitigate climate change, but they have not made a convincing scientific case against human-induced climate change, nor proved that the uncertainties are so great that it is unnecessary to take action now.

Conclusions

This chapter has drawn attention to the following evidence:

• Concentrations of carbon dioxide in the atmosphere have risen by a third since the start of the Industrial Revolution. This has been caused by human activities, primarily burning fossil fuels. The use of fossil-fuel energy and the global emissions of carbon dioxide that stem from it are on a quickly rising curve.

- In the UK, each person's contribution to carbon dioxide emissions is two and a half times the world average. The UK has produced 15 per cent of global emissions since 1750.
- The enhanced greenhouse effect has already caused 0.6°C of warming around the world, which has translated to a 1°C rise in temperature in central England. The impacts of climate change on the human and natural environment are already striking – ranging from a worldwide retreat of glaciers, to a marked increase in extreme weather events, to alarming changes in the seasons.
- If no action is taken to reduce greenhouse gas emissions, temperatures by 2100 could rise by up to 6°C – equivalent to the temperature difference between the last Ice Age and now – or even more if the latest models are correct. This would be disastrous for the climate, the environment and the world's population.
- If there are catastrophic and unexpected changes to the climate system, which cannot be ruled out, the future looks even more bleak.
- Implementation of the Kyoto Protocol will not do much to reduce the risk of further climate change.
- To maintain a reasonably safe limit on carbon dioxide emissions in the atmosphere, the UK will have to reduce its emissions by between 60 and 80 per cent by 2050, as part of a global agreement.

Current trends in emissions, however, are on the worst possible path, risking atmospheric carbon dioxide concentrations of 950 ppm by 2100, with associated temperature rises of *at least* 6°C. The action that has been taken to date is very minor in relation to the scale of the potential catastrophe, despite the fact that the severe impacts of rising greenhouse gases are increasingly understood.

Climate change represents the most serious global problem imaginable, affecting the planet for tens of generations to come. We are now beginning to witness the consequences. Historic and current emissions of greenhouse gases have already committed the world to rising temperatures and sea levels for hundreds of years into the future. Disturbingly, this will be our legacy to the future. Radical and urgent action is needed to ensure that we are not responsible for making much of the planet uninhabitable.

3. As If There's No Tomorrow
Energy Use: Past, Present and Future

Human beings are increasingly voracious users of energy. We have probably used more energy in the last century than in the preceding one hundred centuries put together. Most of this energy comes from burning fossil fuels. Developed countries in particular depend very largely on these for their energy needs. The UK is typical of this, deriving 90 per cent of its total energy requirement from them. It is this fact which makes drastic curtailment of their use to prevent disastrous climate change such a challenge.

What follows is an attempt to trace the rise in the consumption of energy worldwide over the past thirty years. As this book is concerned with how we as individuals can make a difference to climate change, the focus is very much on energy used in the home and for transport. There is then a more detailed look at the UK's energy 'profile' in the past, present and future. This highlights the fact that all the trends point to an upward spiral of energy use – with the related impacts of increased carbon dioxide emissions. Our new, energy-rich lifestyles are destroying the planet.

World energy use
Past and present

World energy demand is increasing. The best measure of this is the demand for what is called 'primary energy'. Primary energy is the measure of the total consumed, including that used or lost in the process of producing energy for the end user – the consumer. Demand for primary energy has increased by 80 per cent from 5,000 million tonnes of oil equivalent (mtoe) in 1970 to 9,100 mtoe in 2001 (see Figure 4 below). Oil remains the largest single source of energy, accounting for over one-third of the total, followed by gas and coal, each of which represents a further 25 per cent. Gas has become relatively more important as a global fuel since 1970. The role of nuclear and hydroelectricity has also increased during this period.

Most of the increase in consumption since 1970 is accounted for by Asia, where it has more than tripled, and North America (USA, Canada and Mexico) where it has increased by around 50 per cent. For the EU countries, it has increased by 40 per cent, mainly due to rising demand in France, Spain and Italy. Despite these changes, North America has remained the region with the highest share of energy consumption, and Africa is still by far the lowest consuming region. Regional shares in 2001 were:

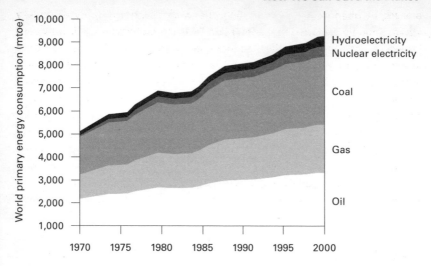

Figure 4: World primary energy consumption, 1970–2001

- **North America:** 29 per cent.
- **Asia:** 26 per cent.
- **EU:** 16 per cent.
- **Non-EU Europe:** 5 per cent.
- **Africa:** 3 per cent.
- **Rest of the world:** 21 per cent.

Like the carbon dioxide emissions it causes, energy usage per person is very unequally distributed across the world. These differences are highlighted when we look at primary energy use for different countries: in 2001, the figure for the USA was 7.6 tonnes of oil equivalent (toe), the UK was half that at 3.8, China was much lower at 0.6, with Bangladesh accounting only for 0.09, little more than a hundredth that of the average American.

Why are we using more energy?

Why is world energy use rising? The simple explanation is that the world population is increasing and many of us are getting richer. Higher incomes lead to higher standards of living and, with that, more production and consumption of goods and services, most of which are underpinned by fossil-fuel energy. Multiply these two factors together and the result is a rapidly increasing demand for energy.

In reality, the relationship between energy use and changes in the economy and population is more complex. Fast economic and population growth are recent phenomena. Before 1500, the world economy grew extremely slowly, mainly because the population did not increase significantly and improvements in productive technologies were small. It was not until around 1820 that economic growth was sufficiently rapid to lead to rising income levels. This was made possible by new technologies and systems of economic organization that were in turn enabled by the wider availability of energy. Fossil fuels were at the heart of these changes. The Industrial Revolution was first powered by coal and, later in the nineteenth century, by oil and gas. The economic and population growth of the last two centuries could not have occurred without these newly identified sources of fossil fuel. In other words, the increased availability of fossil-fuel energy is a cause as well as a consequence of a rising population and increasing economic activity.

Looking ahead

The world's population, economic wealth and energy use are all predicted to continue to rise into the future. Over the next fifty to a hundred years, growth is expected to be strongest in developing countries where, according to the World Energy Council, energy use will come to dominate world consumption, from a position of accounting for about one-third of the total in 1990 to almost two-thirds in 2050. Trends in population growth, economic growth and energy use are discussed below, as are the implications for climate change.

Population growth

World population stood at 6.1 billion in mid-2000 and is currently growing at an annual rate of 1.2 per cent – an extra 77 million people each year. The mid-range estimate for world population in 2050 is 9.3 billion, representing a rise of 50 per cent (the lowest projection is 7.9 billion and the highest 10.9 billion). Growth will occur predominantly in the developing countries. The populations of those that are more developed, currently 1.2 billion, are not expected to change very much because fertility rates are expected to remain below replacement levels.

Wealth and economic growth

Economic forecasts up to 2100 made by the IPCC all show huge increases in gross domestic product (GDP). The forecast global figure ranges from eleven- to 26-fold by 2100, depending on the scenario. For all scenarios, the income differential per person between developed and developing countries is expected to fall considerably from today's very unequal ratio of 16:1 to between 2:1 and 4:1 by the end of the century. This depends on much higher rates of economic growth in developing countries, which are already occurring. For example, the economy of China, with 1.3 billion people, has been growing at around 8 per cent per year since 1980 and is planned to expand four-fold within twenty years. Indeed, between 1990 and 2100, individual incomes are expected to rise three- to eight-fold in industrialized countries but twelve- to 74-fold in developing countries. These levels of growth are dependent on large increases in the use of fossil fuels and other natural resources.

Although this book is concerned with how individuals can make a difference to climate change by addressing the problems of excessive fossil-fuel consumption, it is worthwhile taking a look at the effects of the consumption of other natural resources on such an alarming scale. Many of those we depend on, such as land, fish stocks and fresh water, are reaching the limits of exploitation. There is a measure called the 'ecological footprint' which shows the combined impact of consumption of energy and material resources. The application of this measure reveals that if everyone on the planet consumed as much as each person in the UK, we would need three planets from which to produce the resources and deal with the waste. This is clearly impossible. The GDP predictions for 2100 assume that it will be possible for everyone in the world to consume several times more than current UK levels. The finite nature of the planet's resources and its environmental limits make the prediction both fanciful and dangerous because this level of consumption is obviously unsustainable. Economic growth cannot continue to be pursued as if there were no limits on the use of resources and disposal of waste. Climate change is the most serious issue but it is not the only one.

Energy use

As already mentioned in Chapter 2, IPCC projections are that global primary energy use could rise by one and a half times by the end of this century but, if current trends of substantial economic growth continue, it is likely to increase four-fold. The implications of this for climate change have already been discussed. It clearly cannot be allowed to happen.

The UK energy profile
Past and present

Energy use in the UK has also risen over the past thirty years, but less dramatically than for the world as a whole. It increased by 13 per cent between

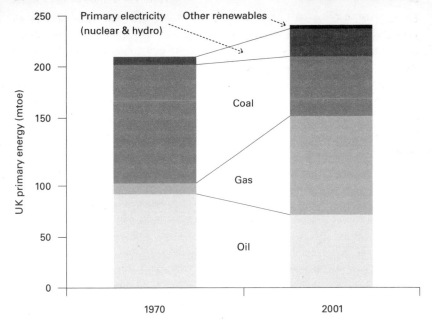

Figure 5: UK primary energy demand, 1970 and 2001

1970 and 2001. As Figure 5 shows (see above), during that period there is a considerable difference in the types of fuel used. Coal and oil dominated in 1970 whereas, by 2001, gas was the single most important fuel. In addition, the output of nuclear electricity has increased as has the contribution from renewable sources, though to a considerably lesser extent. The change in the mix of fuels has largely stemmed from the change in the fuels used in electricity generation and for heating; in each case there has been a move away from coal and oil towards gas. The fuel used for transport remains unchanged as there are no real alternatives to oil at present.

Due to their chemical structures, coal, oil and gas each emits a different amount of carbon dioxide per unit of energy produced, measured in kilowatt-hours (kWh). Coal is the most 'carbon-intensive' per unit of energy at 0.08 kgC/kWh, followed by oil at 0.07 kgC/kWh and then gas at 0.05 kgC/kWh. So to reduce carbon dioxide emissions, gas is the preferable fuel as its use leads to less than two-thirds of the emissions from coal.

However, it is important to note that the rising popularity of gas in the UK has not been in response to climate change concerns. The fact that it emits less carbon dioxide than coal is fortuitous. It is due to a combination of its greater availability since its discovery in the North Sea (and in other areas of the world), its low prices on the open market, and its ease of usage and transportation compared with coal and oil. In addition, it has been favoured for use in power stations. In this case, it is due to the greater efficiency with which gas can generate electricity, and because gas power stations are much cheaper and quicker to build, factors which were soon recognized when the electricity supply industry was privatized and liberalized.

Due to these changes and the increased use of nuclear power and renewable energy, the 'carbon intensity' of electricity – that is, the amount of carbon dioxide emitted per unit of energy – has fallen by 60 per cent since 1970, and now stands at around 0.12 kgC/kWh. As a result, carbon dioxide emissions from burning fossil fuels in the UK have dropped in the last few decades despite increased energy use. In 1970, emissions were 186 MtC, in 1990 they were 165 MtC, and in 2002 they were 150 MtC – that is, almost one-fifth lower than they were in 1970. Thus the UK looks set to be able to meet its Kyoto reduction target for 2010 because, in spite of rising energy demand, it has been able to achieve reductions in carbon dioxide emissions, and this lucky outcome has encouraged the UK to play a leading role in climate change negotiations.

But this is not the whole story. These figures for carbon dioxide emissions do not include the contribution of international aviation because there is no international agreement on how to incorporate them in national inventories. Those from the UK were estimated at 8 MtC in 2002, which would add another 5 per cent to the UK's total. However, aircraft emissions add three times more powerfully to the greenhouse effect than the carbon dioxide component alone, making their effective addition to carbon emissions 24 MtC, equivalent to an extra 15 per cent on the UK's current total emissions. This represents a significant increase on the official figures, taking 2002 emissions to 174 MtC equivalent. Given the considerable growth in air travel in recent years, the addition of aircraft emissions results in the UK's emissions being approximately level, and not falling since 1990. The fact that aircraft emissions are not included in the Kyoto treaty does not mean that they are not contributing to damaging the atmosphere. Indeed, as the fastest-growing source of greenhouse gases, they clearly should be included in government figures *as a matter of urgency.*

Who uses the energy and who emits most carbon dioxide?

We have seen how much energy is used in the UK compared to the rest of the world. But who is using it and what are the main processes leading to emissions of carbon dioxide from it?

The measure for assessing how much energy is used is called 'final energy'. This is the energy actually received by the final user (such as motorists when they buy petrol or householders when they use electricity). It differs from primary energy because it excludes what is lost in getting energy to consumers in the form they want. Figures for primary energy are required to calculate carbon dioxide emissions, but figures for final energy are more useful in understanding where energy is used in the economy. From this point onwards in this chapter, all figures relate to final energy, unless otherwise stated.

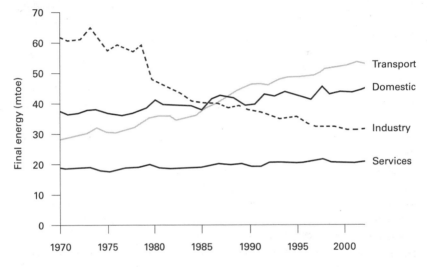

Figure 6: Final energy demand in the UK, 1970–2002

The proportion of final energy used by different sectors of the economy has changed significantly since 1970 (see Figure 6 above). Energy use by industry has almost halved, but use in both the transport and domestic sectors has grown to the extent that transport, which includes freight as well as personal transport, is now the largest user. The proportion of final energy use by the different sectors in 2002 was:

- **Transport:** 35 per cent.
- **Domestic:** 30 per cent.
- **Industry:** 22 per cent.
- **Services:** 13 per cent.

Within these figures, the individual's use of energy has grown considerably since 1970. The UK Department of Trade and Industry (DTI) has estimated that, of total *transport* energy use, 60 per cent is now used by individuals, 24 per cent by industry (mainly for freight), and 16 per cent by the service sector. To get an overall percentage for individuals in the UK, we can take the 30 per cent (domestic use) and add 60 per cent of the 35 per cent used in the transport sector. The total becomes 51 per cent, up from its level of 37 per cent in 1970.

Because of the different fuel types used in each sector, carbon dioxide emissions are not in exactly the same proportions as energy use. In 2000, industry and the domestic sector were each responsible for 28 per cent of emissions, transport for 26 per cent and the service sector for 18 per cent. So, emissions produced directly by individuals were about 45 per cent of the national total.

How have we got to where we are?

We are becoming an increasingly energy-intensive society, often without being aware of it. Very few people indeed know the link between their fuel-based activities and the greenhouse emissions stemming from them. Understanding the powerful forces driving this unsustainable growth in all sectors of the UK economy – transport, domestic, industrial and service – is the first step in reversing it. Population increase is not one of these forces. In common with most developed countries, the UK population has risen fairly slowly, from 55.8 million in 1970 to 58.7 million in 2000, an increase of just 5 per cent. However, the number of households is 30 per cent higher. The effect of this on domestic energy consumption is significant.

Transport

Increasing transport activity is a key feature of modern society. Associated energy use and carbon dioxide emissions have grown faster than in any other sector, with fuel use almost doubling between 1970 and 2000. This has been caused by people travelling much further each year. According to the National Travel Survey, which excludes international travel, each person now travels on average 60 per cent more miles than thirty years ago. The annual total would be considerably higher if air passenger mileage were included. This is the result of far more energy-intensive forms of fuel-based transport by road, rail and air, such as cars, being used, over longer distances and at higher speeds.

On average, cars generate the highest carbon dioxide emissions per passenger for a given journey on land, while the emissions from rail and bus are somewhat lower. However, average figures hide many important variations for individual trips (discussed in more detail in Chapter 8), and comparison between different methods of transport is affected by the number of people in both cars and buses and trains, the energy efficiency of the vehicles, the speed at which they travel and the type of fuel used.

Car travel

Travel by car dominates transport in the UK to the extent that it now accounts for 83 per cent of the distance travelled. This is an increase since the early 1970s when overall travel per person was 7,200 kilometres (4,500 miles) and about two-thirds of it was by car.

In 1970, just over half of UK households owned a car, but by 2000 almost three-quarters had at least one car, and more than a quarter owned two or more. Combined with the increased number of households overall in the country, this has meant that the number of cars has more than doubled, from 11.5 million in 1970 to 23.2 million in 2000. Over half a million new additional cars are purchased annually (that is, not including replacement cars). If just parked end-to-end, these cars alone would require a 3,200-kilometre (2,000-mile) stretch of road, equivalent to a five-lane motorway between Bristol and Edinburgh.

During the same period, car travel has become considerably cheaper in comparison with public transport. Its cost in real terms (including purchase, maintenance, petrol and oil, and tax and insurance) has been almost unchanged since 1974, whereas public transport fares have risen by over 50 per cent. As well as lower comparative costs, the rise in disposable income of around 80 per cent since 1974 means the cost of travelling one mile by car has accounted for a steadily declining proportion of the household budget.

The increasing spread of car travel has displaced other modes of transport. It has been estimated that for each additional car purchased, 300 fewer journeys are made each year by public transport. Not only that, increased car traffic has created a more hostile environment for journeys on foot or cycle, leading more people to abandon these non-motorized methods of travel. The car has largely outcompeted its transport rivals: outside London, three-quarters of commuters travel by car. Furthermore, it has encouraged changes in land use and planning in its favour. Changes in the location of facilities and use of land to better accommodate the convenience of car users now make retreating from car-dependent lifestyles difficult. The consequent increase in the distances that have to be travelled to reach retail outlets, leisure destinations and so on, combined with the inability of public transport to match the door-to-door convenience of the car, have resulted in those without a car facing greater difficulty in accessing what were local facilities. Again this adds to the pressure to buy into the car-dependent culture. In short, the increase in car travel has become self-perpetuating, encouraging people to travel further afield. At the same time, it has locked them into an ever more energy- and carbon-intensive way of life.

Public transport

Much of public transport has been in decline for many years. Since 1970, the average distance travelled per person by bus and coach has fallen by a third. Indeed, owing to the huge increase in car use, whereas in 1970 travel by bus and coach (combined) and by rail accounted for 15 per cent and 9 per cent of passenger kilometres respectively, by 2000 the proportions had both reduced to only about 6 per cent. Bucking the trend somewhat, rail travel has increased from the late 1990s and bus journeys in London are rising. But, in spite of massive public subsidy, the general picture of public transport in the UK could hardly be described as healthy. The reasons for its declining popularity are largely the same as those which have led to increased car ownership and usage. As soon as people have been able to afford private transport, they have generally found its qualities far preferable.

Despite public transport's singular lack of success, it is still commonly seen to be the main means of dealing with transport problems in the UK, particularly those associated with road congestion. The aim has therefore been to increase the number of journeys made by it. However, given its current low share of travel, even an unlikely doubling of public transport use in the future would not in itself reduce car travel very much. Surveys of the changes that have occurred following major improvements in public transport show that these bring about a relatively small amount of transfer from journeys previously made by car. Like new roads, new public transport services create new demands for travel. Relying on public transport to resolve the problem of increasing car travel is unrealistic.

Non-motorized modes of transport

Walking and cycling have many undisputed advantages. They are good for health (regular adult cyclists have a level of fitness equivalent to that of somebody ten years younger), they do not damage the environment, they generate zero emissions, and are often the fastest method of movement for short- to medium-length journeys. Nevertheless, against a background of increasing personal travel, people are walking and cycling less. Cycling dropped massively between 1950 and 1970, with the total distance travelled decreasing by a factor of four. More recently, the fall has levelled off. Nevertheless, the *annual* average is now a paltry 62 kilometres (39 miles), accounting for less than 1 per cent of the distances travelled by car. Walking fell by 20 per cent during the 1990s, with the result that the average annual distance travelled on foot in 2000 was just 302 kilometres (189 miles). This represents less than 3 per cent of total distance travelled and less than fifteen minutes a day spent walking.

The decline of walking and cycling is not simply due to people wishing to avoid physical effort or a response to the wider availability of cars. It is also the result of a transport infrastructure which has been designed to cater for motorized transport users and of land use changes which have increased the typical distances that people have to travel. The consequence is that, intentionally or accidentally, people have been discouraged from using these non-motorized and healthy ways of travelling.

Air travel

International and domestic air travel in nearly all countries is of particular concern for four reasons: first, it is highly energy-intensive; second, its speed enables and encourages people to travel long distances; third, its use is increasing rapidly; and finally, its contribution to global warming is around three times greater than is indicated by the carbon dioxide emissions of a single flight. The additional damage occurs because other greenhouse gases are emitted from aircraft and the fact that this occurs more destructively in the upper atmosphere. These four factors combined have led to aviation representing by far the fastest growing source of greenhouse gas emissions – *far* higher than for any other form of transport.

In the UK, air travel, particularly on international flights, has risen sharply. Passenger kilometres flown from airports more than doubled between 1990 and 2000, from 120 to 260 billion. This is the equivalent of each person flying over 4,400 kilometres (2,750 miles) per year – from London to Paris and back six times. As with car travel, increased air travel is associated with increased personal wealth and its availability at a low price. This is no accident: aviation has a privileged position in the economy, as discussed in Chapter 5.

The demand for increasing international travel is primarily explained by the attractions of widening opportunities for leisure and tourism. Only 14 per cent of visits abroad by UK residents in 2001 were for business. Travel and tourism is now the world's largest industry, accounting for 11 per cent of the world's GDP. Economies in many countries are becoming increasingly dependent on foreign earnings from visitors. Almost without exception, the ecological implications to the planet of tourism based on air travel have been overlooked. This wilful 'turning a blind eye' has elements of black comedy. For example, a writer in a UK newspaper recently advised, with no apparent sense of irony: 'Go to the Maldives if you want to see it before global warming swamps the islands.'

In addition to ever more distant holiday locations being promoted, there is an increasing number of attractions and activities which are designed to draw on a wide, international audience. These range from sporting events to environmental conferences. The fact that even environmental organizations arrange such events shows how deeply embedded they are in our current culture. The World Wildlife Fund, for instance, holds conferences each year in different locations around the world: 1998 Interlaken, 1999 Indonesia, 2000 Katmandu. The Global Justice Movement attracted 40,000 members to the World Social Forum in Porto Allegre in Brazil in 2002.

Non-governmental organizations are not exceptional in this regard; they are simply responding to the same pressures and drives as businesses and governments in convening larger and more international meetings and events at increasing frequency. This globalization of many types of activity is not sustainable, however, depending as it does on a considerable amount of flying. A reduction in international air travel is essential. Planting trees to absorb carbon dioxide from the atmosphere and 'offset' the emissions from flying, as has been suggested, is not an adequate response (the limitations of tree planting are explained in Chapter 6). Even if there were no alternative ways of holding these events – and, technological breakthroughs such as satellite link-ups and teleconferencing suggest that there are – the impact of this type of activity can only rarely be justified.

Why are people travelling more?

People are travelling much longer distances than in the past partly because they can now afford to do so and partly because they have to. Improvements in the quality of transport have enabled this. Within the UK, car transport has become ever cheaper and more widely available. Provision of the supporting infrastructure of new roads, bypasses and motorway widening has been a priority of successive governments, and this has been delivered to the driver free at the point of use. Distant holiday destinations made more accessible and competition in the tourist industry, combined with more air routes and cheap fares, rapid rail services, cruise liners and coaches, have led to a spectacular widening of leisure destinations. More and more people are able to travel further, faster and at less cost, and can treat the world as their oyster. This growth of opportunities has created a vicious circle – the higher the speed that people can travel and the more distant the available destinations, the greater becomes the demand for transport and the more damaging are the impacts on the environment.

Not all additional travel is voluntarily chosen. As travel possibilities in-crease, so does the obligation to travel further. The rise in personal mileage in the UK is due to increased journey lengths, not to more journeys being made. People are having to travel further to get to where they want to be because of the decline in local services and facilities and because they are choosing jobs which entail commuting over longer distances from home than previously.

During the last century, when progress was considered synonymous with reducing the obstacles associated with time and distance, faster travel to more distant destinations was, and still remains, the goal. This will have to change. In the absence of a miraculous technical fix (the poor prospects for which are discussed in Chapter 6), travel will have to become more local, less frequent, less energy-intensive and slower. Avoidance of transportation is at the heart of the transition required. We need a less mobile society, which is the reverse of the one that the public wish for and which politicians and industry are planning for.

Freight transport

Total freight movement between 1970 and 2000 rose by over 80 per cent, echoing the trend for more personal travel. Most of the increase has been in road transport. About two-thirds of freight (in tonne-kilometres) is now moved by road, a quarter by water, and the remainder by rail and pipeline. For freight, the key factor is distance travelled. Longer journeys account for most of the increase, with only a minority due to a greater amount of goods being carried. The increase in the average length of journey results from changes in distribution patterns, such as the development of regional rather than local distribution centres, and 'just-in-time' deliveries. These evolutions in business practice have come about because of the cheapness of road transport but they are detrimental to both the local and global environment. Strikingly, a considerable amount of the fuel used by the transport system is used to carry fuel itself to the customers who use it. Nearly a third of total tonne-kilometres was dedicated to moving coke, coal and especially petroleum products, mainly by water.

Air freight is the most polluting form of transport and is increasing rapidly. The tonnage carried by air, landing or taking off at UK airports, rose by about 9 per cent each year during the last decade and the trend is continuing. Originally, air freight was used only as a means of filling excess storage space in passenger aircraft, but as wider-bodied planes were developed, freight transport became an important activity in its own right. Now about a third of air freight landing or taking off from UK airports is carried in all-cargo aircraft. Thus, more freight is being moved, both on longer journeys within the UK and to and from abroad by air. From a climate change perspective, these trends are unsustainable and must be reversed.

Household energy use

Household energy use amounts to 30 per cent of the UK total and is responsible for 28 per cent of the country's carbon dioxide emissions. Space heating now accounts for around three-fifths of this energy use, water heating for one-fifth, and cooking, lights and appliances for the remaining fifth. Ninety per cent of households now have central heating, four in five using gas.

Demand has been growing steadily over recent decades and continues to grow annually at around 1 per cent. Several factors account for this. First, electricity is now being used for purposes which were not even invented ten, twenty or thirty years ago – for microwaves, personal computers, CDs, DVD players, digital television and so on. Second, energy prices have fallen relative to incomes: fuel bills now account for only 3 per cent of the average household budget compared with 6 per cent in 1970. Despite this, there are still millions of households in 'fuel poverty' – needing to spend at least one-tenth of their income to afford adequate energy services. Fuel-poor households often cannot afford to spend this amount of their income on energy, and so suffer from cold homes and associated medical problems. Nevertheless, in general, the opportunities for using energy in the home have expanded at the same time as the cost of doing so has decreased.

How has energy use been changing?

Energy use in the average home has changed little since 1970; at the start of the twenty-first century, it was similar to that of three decades before, despite warmer homes and more electrical equipment. This is due to the considerable improvements in the efficiency with which we heat our homes – with wasted heat from inefficient buildings and inefficient boilers falling significantly. Over the same period, electricity use in lights and appliances increased by over a third. However, because the number of households has risen by a third, total energy use in the domestic sector has risen to a similar extent.

The rise of equipment ownership and usage

The amount of energy used in the home is determined by a combination of ownership levels of equipment, its frequency of usage, and the quality of its technology. While the ownership of lighting and other electrical appliances has shown particularly strong growth, there have also been considerable changes in how energy is used for space and water heating. The ownership of central heating has risen from about a third in the early 1970s to over 90 per cent by 2000, and that of washing machines from two-thirds also to over 90 per cent. Microwave ovens and home computers, neither of which existed in 1970, are now owned by 85 per cent and 50 per cent respectively of UK households. Ownership of other new products, such as digital televisions and digital decoder boxes, is also increasing rapidly.

In the past, much of the UK's housing was poorly heated and draughty. However, with more people having central heating and better-insulated homes, higher levels of comfort and convenience have been obtained. A combination of a desire for increased temperatures and the capacity to achieve them, both technically (through improved efficiency) and financially, has led to dramatic changes. Most of the benefits have entailed additional energy use. The most striking example is the increase of internal temperatures during the heating season. In 1970, the average internal temperature of homes was about 13°C. It has risen steadily since then and is now heading towards a national average of 19°C.

Other changes in the home include more frequent use of hot water for washing machines, tumble dryers, and so on, and increased use of televisions, personal computers and other entertainment equipment. However, in a few cases, changes have led to decreased energy consumption: less electricity is now needed for the lower-temperature wash cycles preferred in washing machines, and microwave ovens use less energy than their conventional rivals.

Social changes, as much as new technologies, have also added to energy demand. One cause is that energy consumption within households has become individualized, with more than one television per household being standard, allowing for different viewing patterns for its members. Bedrooms now tend to be heated to a similar level as living areas, allowing children and adults to pursue different activities elsewhere in the home. This requires additional energy for heating, lighting and use of other equipment.

Rising expectations
The UK has become a wealthier nation and, with it, expectations of normal life have risen. It is no longer considered acceptable to live with the lower temperatures in the home that were normal previously and it is thought quite acceptable to wear light clothing indoors, even in the depths of winter. Having central heating, a fridge and freezer, cooker, microwave oven, a couple of televisions, a VCR or DVD player, a washing machine, numerous radios, music equipment, a personal computer, mobile phone chargers and many other gadgets is taken for granted in most households. It seems likely that, in due course, yet more energy-guzzling technologies (such as air conditioning, outdoor patio heaters, large American-style fridge-freezers, and digital entertainment equipment) will also become commonplace. Improvements in energy efficiency cannot keep up with these ever-rising demands for energy-based equipment. Technology alone, despite its impressive past

performance and huge promise in the building sector, will not be enough to halt the demand for energy in the domestic sector (more details in Chapter 6). Rising expectations, if met, will inevitably result in causing further climate change.

Industrial and service sectors

As explained in Chapter 1, one of the main purposes of this book is to focus on the role of the individual in helping to save the planet. However, it is important to remember that most energy use in the industrial and service sectors is for the benefit of us as individuals. The energy used in these sectors provides the transport systems that we depend on, heats and lights the shops we visit and the schools our children attend, powers the computers of our banks and building societies, and provides the products we buy and the food we eat. As well as reflecting the types of activity undertaken, and the efficiency with which it is carried out, energy consumption in these sectors depends on the level of activity – on how many goods and services are produced and consumed.

Between them, the industrial and service sectors use just over a third of UK final energy, with industry using 22 per cent and the service sector 14 per cent. The largest industrial user of energy is the chemicals sector at 22 per cent, followed by iron, steel and other metals at 14 per cent, metal products, machinery and equipment at 13 per cent, and food, drink and tobacco at 12 per cent. Energy demand in the service sector, which includes the public sector, such as schools, hospitals and local government, is dominated by the energy used in buildings, particularly space heating, air conditioning and hot-water provision. Lighting is its largest single source of electricity demand. As illustrated earlier, industrial use of energy in the UK has been in sharp decline since 1970, levelling off in the mid-1990s, whereas energy use in the service sector has shown slow but steady growth throughout the period. These divergent trends reflect changes in Britain's economic activity, with energy-intensive industry becoming a decreasing part of the economy and the service sector assuming increasing importance.

Because we import goods from abroad, energy is being consumed and carbon dioxide emitted on our behalf in other countries of the world. The same process happens in reverse when goods produced in the UK are exported abroad. We do not know whether the national emissions represent a true indicator of our 'carbon burden', or whether, on balance, other countries are emitting excess carbon dioxide to provide us with goods. However, the loss of energy-intensive industry from the UK, such as steel manufacturing, suggests that imports of energy-intensive goods from abroad exceed exports.

From farm to fork: the hidden cost of food

Our lifestyles are changing and causing an increase in the levels of energy used in areas we are often completely unaware of. The case of food production highlights how ingrained the increased consumption of energy has become in our day-to-day lives. Energy use in the food system has grown considerably. At present, about six calories of fossil-fuel energy are needed to produce one calorie of food. However, if food is imported from a distant location, more than ten times as much energy can be required: for every calorie of carrot flown in from South Africa, 66 calories of fuel are used. The reasons for increasing energy use range from dietary changes to new patterns of retailing, and are explored below. The story of food production demonstrates how, even without realizing it, we as consumers are involved in making choices which are having a further damaging effect, albeit indirect, on the climate.

Trends in food choice and cooking

There has been considerable change in eating habits over the last thirty years, leading to a preference for foods which require more energy for production and distribution. Government data show a significant fall in the consumption of traditional British greens, such as cabbage, kale and sprouts, and root vegetables, such as carrots, turnips, swedes and parsnips. Instead people are buying fresh and frozen imports of more exotic vegetables. Because these have to be transported further, they entail far higher use of energy than food grown in the UK. In addition, diet has become much less affected by the seasons as many fruits and vegetables have become available all year round. While this may be convenient and appreciated by consumers, it imposes an energy penalty as these have to be grown either in heated greenhouses or further afield and imported.

Another influential factor is the expansion of the market for convenience food and ready-made meals, which require more energy in the total process of getting them to the consumer. This trend towards convenience food has been facilitated by two technologies which have developed in parallel: the freezer and the microwave oven. Although the microwave represents a low-energy cooking technology, the use of freezer and fridge space, in transport from the factory, in retail outlets, and in the home, increases the overall energy used for food storage. One has only to compare energy use in this instance with that of the traditional, fuel-free larder.

Growing

Conventional farming, particularly meat production, is increasingly energy-intensive. A key reason is that, in contrast to organic agriculture systems, it relies on the manufacture of artificial fertilizers, mainly nitrogen-based, which uses considerable amounts of fossil fuel. The differences in energy used can be considerable: organic arable production uses one-third less and organic dairy three-quarters less than their conventionally farmed equivalents. Modern, industrialized systems of production, while cost-efficient in commercial terms, often depend heavily on fossil-fuel input. Farming is no exception.

Transporting

The issue of 'food miles' is gaining prominence for good reason. The carriage of food now accounts for over a third of all UK road freight. The distance that food is transported by road has increased by half since the late 1970s, and food now accounts for over 10 per cent of the volume of goods transported by air. A recent calculation showed that a basket of twenty fairly ordinary items from a supermarket, including Chilean wine and American apples, clocked up over 100,000 food miles. Both imported and nationally produced food is travelling further than ever, in more environmentally damaging ways, before it reaches the shops. The UK's foot-and-mouth disease crisis of 2001 highlighted the increased movement of livestock around the country, with some animals being taken on eight different journeys between the farm where they were born to the farm at which they would be reared to maturity.

The transport burden of food does not end at the supermarket. The miles travelled by householders to do their shopping also have to be considered. Shopping now accounts for 13 per cent of all personal mileage per year. In the last thirty years, the average distance travelled for this purpose has more than doubled and the proportion of shopping mileage made by car has risen so steeply that it now accounts for 86 per cent of the total.

Retailing

Patterns of food retailing have also changed considerably during this same period. Nowadays, the vast majority of people do their main shopping at a supermarket, compared with just one-third in 1970. Supermarkets have turned into superstores, largely located out of town. Much of the growth of these retail outlets was facilitated by planning policies and made possible by the growth in car ownership. Once set in motion, the change in patterns of shopping has proved self-reinforcing. Supermarket expansion has contributed to the decline in local shops, leading to greater dependence on the supermarkets, which have themselves consolidated into ever larger and more attractive outlets by offering wider choice and lower prices. It has also enhanced the attractions of using a car, not only to carry the larger quantities and weight of goods now bought at supermarkets but also to cover the greater distance that typically has to be travelled.

Because of the dominance of supermarkets, ways of doing business that suit them influence all the other players in the food chain, including suppliers, farmers, manufacturers and transport companies. Their sheer size and centralized organization make it difficult for them to supply locally produced food, although some change towards local sourcing may be expected, primarily to aid UK farmers rather than for environmental reasons. While the growth of supermarkets has taken place in response to the preferences of most shoppers, it has also played a major part in the intensification of energy use in the more complex process of getting food from the farm to the home.

What does this tell us?

The increase in energy intensity of the food chain over recent decades has not been caused by a single factor. It is the product of many influences, including planning policies, the concentration of retail outlets, changes in production practices, changing consumer tastes, new technologies (such as freezers and microwave ovens) allowing new forms of consumption, declining costs of car and air transport, and increasing demand for a wider choice of goods from around the world. These changes affect many supply chains. In addition, as with other forms of consumption, the energy impacts of food depend on how much is bought. Over-consumption of food in the developed world is reaching epidemic proportions and has a direct, negative effect on personal health and wellbeing. In the UK, over half the adult population is classified as overweight – itself a clear metaphor for the damage to the planet from our excessive use of energy.

The problems in today's food system are complex, but the alternative is simple to describe. In a less energy-intensive system, food would be more locally produced, more seasonal, more organic, less processed and purchased closer to home. Achieving this will take time, thought and effort and will necessarily narrow some aspects of choice. But public policy and consumer practice may well have to be revised to that end.

Future energy use

Unless there are radical changes, life in the UK is expected to become more energy-dependent into the future. Growth in energy use will lead to growth in carbon dioxide emissions, particularly once the trend towards using an increasing proportion of gas in the energy mix has peaked and the contribution of nuclear energy has declined as its power stations go off-stream. Within general growth, there are 'hot spots' of activity which are likely to lead to sharply rising energy usage in some sectors. The following section summarizes current official energy projections to 2020, looks at a variety of scenarios for 2050 and considers their implications.

Government projections to 2020

The DTI produces projections for energy use and carbon dioxide emissions on a regular basis. These are based on analysing historical trends in energy use and their relationship to factors such as economic growth and fuel prices. They also take account of current government policies on energy. Two different fuel-price scenarios (low and high) are generated, and energy demand is then calculated for three variations on the rate of economic growth – low, central and high – resulting in six scenarios. Based on the central scenarios, final energy demand is predicted to rise annually by about 1 per cent up to 2010. These figures include international air transport. Annual growth is expected to be strongest in the transport sector (1.7–1.9 per cent) but the domestic and service sectors also show strong growth at 1 and 1.1 per cent respectively. However, the structural shift in the economy away from heavy industry is expected to continue, giving low growth in this sector. By 2010, the projection is that final demand will be 10 per cent higher than in 2000.

Assuming the same split between freight transport and passenger travel as at present, direct personal consumption will then account for 49 per cent of the total – about the same proportion as today. There are no projections for final energy demand for 2020. However, primary energy, which is expected to grow by 7 per cent to 2010, is predicted to increase by 12 per cent by 2020. Primary energy growth will be lower than final energy because the energy losses in power generation and processes, which are not counted in final energy, are expected to fall considerably.

Carbon dioxide emissions (with the important exception of international aviation) are likely to reduce up to 2010, despite growth in energy demand. This is due largely to the continuing switch to gas in the electricity supply industry. However, emissions are expected to rise between 2010 and 2020: in the central scenarios outlined by the DTI, they will be just 1–3 per cent below 1990 levels by 2020, but on an upward trend. Moreover, because the projections do not include international air travel emissions – as this is not required under the IPCC methodology – these percentages represent a considerable underestimate. In reality, the UK's carbon dioxide emissions will have grown. Based on the government's own figures, carbon dioxide from international aviation could be double what it was in 2002 by 2020, adding

8 MtC to total emissions. As the damage from all greenhouse gases from aircraft is three times that from carbon dioxide alone, emissions from this source would rise by 2020 to 13 per cent of the total.

Longer-term projections

For longer-term energy projections, a different modelling approach is required to encompass a broader range of possible future developments. There are no official projections covering UK energy use and carbon emissions for the next fifty years. However, a 'scenarios' approach, developed by the UK's national Foresight Programme, has been used to help envisage a range of possible futures and the energy to which this would lead (see Figure 7 below).

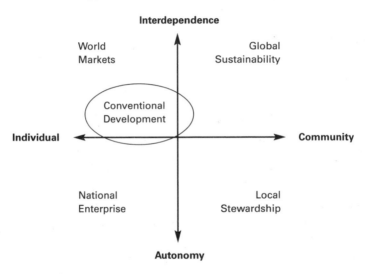

Figure 7: Foresight scenarios for the future of the UK

These Foresight scenarios are based on plausible future changes in social values and governance systems. Social values range from highly individualistic to community-focused ones (such as environment). Highly individualistic societies do not respond to collective environmental concerns and are focused on personal wealth. The global governance systems considered range from, at one extreme, a totally interdependent world, to one in which countries and regions are run on much more autonomous lines than they are today. More interdependence is associated with increased world trade, travel and economic growth. These two dimensions are taken as axes that define four scenarios: World Markets, Global Sustainability, National Enterprise and Local Stewardship. They are quite similar to the scenarios which the IPCC used for their global emissions scenarios and which were discussed in the previous chapter.

World Markets is the scenario most likely to occur if current trends continue. Here national economies and decision-making bodies become ever more locked together, and social values are derived more from the preferences of the individual than from objectives based on the wider public interest. National Enterprise incorporates the same individualistic values as World Markets, but the UK is much less well connected to the rest of the world – a 'fortress UK' scenario. Global Sustainability envisages a globally oriented but community-focused world, resulting in high-tech solutions to environmental problems. Local Stewardship is a 'happy hippies' scenario in which solutions are found for environmental problems within the local community.

These Foresight scenarios were used by a government think-tank to esti-mate UK carbon dioxide emissions in 2050. Emissions were expected to rise from the base year of 2000 by 20 per cent in the World Markets scenario, and by 9 per cent in the National Enterprise one. They showed that a 60 per cent reduction in emissions could be achieved in both the Global Sustainability and the Local Stewardship scenarios, although by very different means. The Global Sustainability scenario requires extensive use of hydrogen and some sequestration of carbon, in addition to major increases in renewable energy use and energy efficiency. As Chapter 6 will show, this is a highly optimistic interpretation of the technical possibilities. The Local Stewardship scenario achieves a 60 per cent reduction owing to greater energy efficiency and wide use of sources of renewable energy, but also lower economic growth and social change.

Thus the highest increase in carbon dioxide emissions is predicted in the scenario that follows current trends, giving an increase of 20 per cent rather than the decrease of 60–80 per cent that is required. The other striking result is that only those scenarios with changed social priorities and the objective of sustainability as a driving force could possibly achieve the required reduction in emissions.

Energy increases: the 'hot spots'

Within the generally predicted rise in energy use, there are 'hot spots' due to the generation of more demand which could lead to energy use that exceeds current predictions. These are increases in the following:

- National and international air travel, possibly at an accelerating rate.
- Car ownership, use and dependence.

- Controlled indoor and outdoor climates, including air conditioning in cars and buildings.
- Number of households.
- The take-up of existing and new uses for electricity.

The rise of national and international air travel

Both national and international air travel are expected to grow strongly, negating many of the 'carbon savings' achieved through government policies in other sectors. The government mid-range forecast is for an annual growth of 4.25 per cent. However, in recent years this has been exceeded. At current rates, four more 'Heathrow' airports would be required by 2020 and nearly six by 2030. In the longer term, without any limiting measures, the government's forecast implies that UK airports will be serving over 1 billion passengers a year by 2050, over five times the 180 million served in 2002. Even a much lower rate of growth has alarming implications: as has been seen, the government expects current carbon dioxide emissions from international aviation to double by 2020, from 8 MtC to between 14 and 16 MtC. Taking into account the added impact upon global warming of these emissions, this amounts to at least an additional 24 MtC by 2020. The effect of this growth by 2010 would be to negate at least half of current government efforts to cut global warming by that date. The disturbing contradiction between government policies to facilitate airport growth and air travel and those to reduce greenhouse gas emissions is discussed in Chapter 5.

There is every reason to believe this predicted growth in air travel if no action is taken to curb demand. All the pressures in our culture are to travel more, to fly further, to be more international. The travel sections of newspapers clearly show the avenues for future growth; they are full of advertisements and articles encouraging ever more flying – weekend shopping breaks to New York, weekend safari trips to Kenya, three nights watching horse racing in Dubai, city breaks in South Africa, day trips to Reykjavik. As one journalist explained: 'Today's travellers are not only short on time but are also on the lookout for instant hits of caché.' There is a near-universal collusion in this dream of ever further and faster travel to exotic locations, promoting the belief that distant travel is horizon-expanding, interesting, high-status, luxurious and something to which we can reasonably aspire and, as incomes rise, enjoy. From this perspective, and as discussed in the next chapter, the fact that it is also intrinsically environmentally highly destructive has to be ignored.

Car-based lifestyles

Car use not only has positive benefits for individuals (personal mobility, prestige, comfort, convenience and cost) but it is also strongly and increasingly reinforced by external forces in society which have placed car-borne accessibility at the centre of land use planning well into the future. Not surprisingly, dependence on it will also increase further: a 20 per cent increase in its use is forecast by 2010, despite a planned £100 billion investment in public transport during the decade.

Increased car use can be expressed as a number of vicious circles:

- More cars → fewer journeys by public transport → reducing revenues to public transport → leading to poorer levels of service → increasing the attractions of the car.
- More cars → more provision for motorists of new roads, parking facilities, congestion-relief schemes → more use of cars.
- More cars → more developments serving car owners → decline of local shops, hospitals and other services and facilities → increasing dependence on the car.
- More cars → more hostile environments for pedestrians and cyclists → further decline of walking and cycling → more use of cars.
- More cars → streets more dangerous for children → leading to parents 'chauffeuring' children (and for more years of their childhood) → more use of cars.

Not only are there social, economic and planning factors that reinforce dependence on cars, there are also psychological factors at work. Research shows that prolonged car use results in greater dependence by users, whose knowledge of and interest in alternatives decline, so that a car-based lifestyle becomes an addiction. Very quickly, cars become converted from 'luxury' to 'necessity'. It is clearly going to be far from easy to reverse dependence on cars, despite the many social and environmental benefits this will bring.

Heating and cooling

The number of locations in which we wish to raise our level of comfort is also increasing, expanding from homes, shops, leisure facilities and places of work, to include cooling in cars (air conditioning) and heating on patios and in gardens. These trends in the rising use of energy for both heating and cooling the spaces we inhabit are resulting in what has been called 'thermal monotony', the same internal climate all year round.

Energy use for cooling epitomizes the process of 'positive feedback', where the response to an issue exacerbates the original problem. As the climate gets warmer, space cooling will be sought in more locations, which will then require further energy, thereby adding to global warming, in turn requiring further cooling and so on, in an upward spiral. Uses for energy-intensive air conditioning equipment are increasing. One estimate suggests that 70 per cent of new cars in the UK now come with air conditioning as standard: the problem is that air conditioning adds around 15 per cent to fuel use. New offices too, with few exceptions, come with air conditioning as standard (and lots of unshaded glazing and inefficient office equipment warming the air, increasing its attractions), and air conditioning is being retro-fitted in older offices. Domestic air-conditioning systems are already being sold in the UK. Indeed, the government expects that domestic air conditioning will be increasingly provided in homes in the future.

Conversely, as the climate warms, less energy should be required for space heating from autumn to late spring. However, in the domestic sector more rooms are being heated to higher temperatures for longer periods and it is not known where these trends will stop. The usual assumption is that average temperatures will continue to rise but will not exceed 21°C. However, average temperatures in Sweden already exceed this, and there is no guarantee the UK will not follow suit. As a rough rule of thumb, each extra 1°C requires 10 per cent more fuel. If everyone chose to heat their homes to 22–23°C, energy consumption for space heating would be 10–20 per cent higher than is currently being predicted, and 30–40 per cent higher than today (with temperatures at around 19°C), all other things being equal. Preference for more control of the internal climate is one of the factors leading to increasing disruption of the external climate!

Number of households
Compared with 2000, the government expects there to be just over 4 million additional people in the UK by 2050, but almost 3.5 million extra households. The decline in average household size is directly linked to women having fewer children and the age structure of the population changing as a consequence of lower fertility and people living longer. In addition, differentiation in living arrangements and changes in marriage behaviour also result in smaller households. By 2050, the population is expected to have levelled out at 66 million, but the number of households will still be rising as more people choose to or have to live on their own. It is this high growth rate which makes a focus on the number of households important.

Both the number and size of households influence total energy consumption in the domestic energy sector. Household size is important because the same population living in smaller households uses more energy than if living in larger households. Per person, energy consumption in smaller households –

those containing fewer people – is much greater than for people living in larger ones: the 'fixed costs' of energy, like heating and lighting, having a fridge, and watching television, are shared between fewer people. Additional households also require their own products, such as homes and cars, which leads to energy demand in the industrial sector that would not have been created if people lived in larger households.

Greedy gadgets

Electricity is a versatile fuel and its use is rapidly increasing, particularly in the domestic and service sectors. It is the most carbon-intensive form of energy available to consumers; each kilowatt hour of electricity results in over twice the carbon dioxide emissions of a kilowatt hour of gas. This is why an increase in electricity usage is particularly serious.

New uses for electricity are easily come by in the digital age. Oxford University calculated that digital TV receiver/decoders alone could add 5 per cent to domestic electricity consumption by 2020. Electricity is increasingly used for household gadgets which are on 'stand-by' mode – doing nothing but waiting to be used. Research shows householders are often unaware that their gadgets are using electricity just by being plugged into the wall. Indeed, around 10 per cent of household electricity is used in this 'invisible' way. Music systems are now sold without a true off-switch. The only way to stop them using electricity is to pull the plug out of its socket. Many of these new gadgets use only a small amount of electricity individually, but national consumption is considerable. In total, electricity use from domestic lights, appliances and water heating is expected to grow by about 12 per cent by 2020 – without taking account of future electrical gadgets not yet invented. There is also considerable concern about the increasing electricity demand in the commercial sector from internet 'hubs' (groups of computers and other equipment) and mobile phone masts. Internet usage and mobile phone ownership are two of the fastest growth areas in commercial activity.

Limits to growth in energy use?

If trends continue as they have been doing, there are few limits to energy growth on the horizon. Spectacular success has been achieved in finding new ways of using energy, especially for electricity in both the household and business sectors, and there will certainly be new, 'essential' items available in the future. People can afford to use more energy and thereby enjoy improvements in their quality of life as they see it. When roads become crowded, the government builds more for the additional traffic. As airports become more crowded, the answer is seen as building more airport capacity.

New power stations are constructed in an effort to ensure 'security of supply' – so that the lights never go out even from exceptional demand for short periods in the winter. There seems to be little acknowledgement of the environmental predicament: no parallel action is being taken to ensure 'security of climate'. The response of individuals, business and government is discussed in Chapters 4 and 5.

Conclusions

This chapter has shown just how deeply embedded the use of fossil-fuel energy is in all aspects of our lives. The first important step that must be taken is to face the global and national evidence of the consequences. Key to this is to acknowledge the fact that the fall in UK emissions of carbon dioxide since 1970 has largely been a lucky historical accident rather than the result of an underlying transformation of the UK economy towards a sustainable future. It is also explained by the fact that the calculations do not include international air travel.

The key points highlighted in the chapter are:

- Use of fossil-fuel energy is rising fast worldwide and, without highly effective political intervention, will continue to do so.
- Energy use per person is very unevenly distributed across the world, as are the resultant carbon dioxide emissions. In sharp contrast to individuals in affluent countries, those in many developing countries currently use very small amounts of fossil fuel.

In the UK:

- Primary energy use has increased by 13 per cent since 1970 but, despite this, UK carbon dioxide emissions from burning fossil fuels have fallen by about a fifth because of the switch to less carbon-intensive fuels.
- At present, calculations of carbon dioxide emissions exclude international air travel as these are not incorporated in current statistics. This gives a false picture of what has been achieved. When emissions from international air travel are included in the national inventory, as they must be, UK emissions can be seen to be steady rather than falling.
- Half the energy consumed is directly used by individuals, a proportion that has increased over time.
- Energy use for transport purposes has grown rapidly due to rising dependence on cars, far more air travel and considerable growth in the distance travelled for all types of journey.

- Household energy use has risen steadily, primarily due to an increase in the number of households and more use of electricity for lighting and appliances in the home.
- Industrial energy use has fallen, due to a shift away from energy-intensive manufacturing, while energy use in the service sector has increased slowly.
- In the absence of the adoption by government of severely restrictive policies to deter energy-intensive activities, energy use in the UK will continue to rise for the foreseeable future. After 2010, carbon dioxide emissions will begin to rise too.

In summary, energy use and carbon dioxide emissions are predicted to continue increasing. In common with other countries, particularly in the developed world, the UK is heading inexorably in the wrong direction: it is moving away from a path that would lead to a significantly lower energy economy and to drastically reduced carbon dioxide emissions. This cannot be allowed to happen.

4. Excuses, Excuses, Excuses
Personal Responses to the Prospect of Climate Change

The previous chapters have shown that climate change is the most serious threat facing humankind, and have suggested (and future chapters will further prove) that our current response is totally inadequate. The way we are living and planning to live is taking us inexorably to a climatic Armageddon. Given this, why has so little been done both at an individual and collective level?

One of the answers to this question is that people have managed to come up with excuses for not taking action. A major part of the reason for this is that in the UK population there is widespread public ignorance of, or denial about, the causes and dramatic effects that climate change is already having on our environment. An Energy Saving Trust report in 2003 on public attitudes to and understanding of climate change, recorded an astounding 85 per cent believing that they will not witness the effects for decades.

Excuses are widely deployed in everyday conversation, in professional debates, and by columnists in newspapers and commentators on television and radio. Superficially, they can seem reasonable. In fact, they are insidious, wrong-headed and dangerous in that they discourage people from facing up to the realities of climate change. In order to take effective action to safeguard our future, it is important to be able to identify and challenge these excuses and move beyond them. This short chapter identifies the ten most common excuses used, looks at their psychological origins, and comments on them.

Why do people need excuses?

If we sincerely face up to what we should do to prevent predictions of climate change coming true, some uncomfortable lifestyle changes must be made. The clearest example is the need to cut down steadily on air travel, perhaps restricting it eventually for use only in exceptional circumstances. This is not a welcome prospect: it indicates that in order to secure our own long-term interests and those of future generations, it will be necessary to make unwelcome alterations to our current ways of life and aspirations for the future. There will, understandably, be a great deal of resistance to this message, which is already reflected in the wide range of excuses for inaction that are put forward and, more commonly, in the avoidance of discussion about even the *implications* of climate change.

A psychological approach can help explain how it is that many of the excuses have come to be accepted as valid and necessary. Coping in situations of conflict, in this case between immediate wishes and long-term interests, relies on employing well-known psychological devices to maintain a mental equilibrium. These defence mechanisms are largely derived from the unconscious and include repression, suppression, denial, projection and dissociation. They provide a shield against the generation of anxiety and self-directed appeals to conscience in the face of unpalatable truths. As they are generally not consciously employed, it can be difficult for individuals to recognize their origins. As short-term measures, their use is understandable and may represent necessary components for coping with the real world. In the longer term, they can be seriously detrimental both to the individual and, in this case, to the future of the planet and the generations succeeding us.

The top ten excuses

Ten of the most common excuses for failing to respond to the need for climate change are listed below, along with a questioning of their validity. The types of excuses in the boxes are edited versions of a miscellany taken from newspapers, journals, business brochures and discussions on relevant issues.

Excuse 1: I don't believe in climate change

The significance of climate change is often denied by doubting its scientific basis. Individuals find it hard to deal with the contradictions and incoherence that they may have encountered in reports about global warming. Influenced by 'sceptics', the media sometimes present climate change as theory rather than reality, and salient bits of information are picked out by individuals which allow them to disregard its implications by claiming that climate change is either unproven or would actually benefit the UK. In the circumstances, it is therefore not surprising that the Energy Saving Trust report recorded such a disturbingly high proportion of the population in denial about climate change.

- No one knows for sure about global warming.
- Thomas Malthus got it wrong in 1798 in calling for population control on the grounds that the world's resources were running out.
- 'Limits to Growth' (the Club of Rome Report, 1972) was wrong in its claim that, based on growth rates at the time, many world resources would be exhausted by 2000.
- The climate isn't really getting warmer, there were vineyards in England in the Middle Ages.
- I am looking forward to living on the Sussex 'Riviera'!

These responses are based on the psychological mechanisms of 'denial' and 'repression/suppression'. Denial is the process of refusing to acknowledge certain aspects of reality, choosing to disregard something because it is either painful or difficult to face up to. It involves a considerable amount of self-deception and distortion of evidence, enabling preferred action to continue in the face of conflict, distress and uncertainty. It is reflected, for instance, in people not wishing to be informed about the link between flying and its damaging contribution to climate change. The phrases 'burying heads in the sand' and 'turning a blind eye' describe the related defence mechanism of repression. There is the subconscious hope that 'it will go away' if we ignore it. The more conscious version of repression is 'suppression'. Information is deliberately shut out – as in refusing to read newspaper reports or even literally 'blocking our ears'. These powerful, largely subconscious responses allow people to continue to wheel out 'excuses' which would not be credible if measured up against the actual scientific evidence for climate change.

Response

As explained in Chapter 2, the world's scientific community is almost united in agreeing that human-induced climate change is a reality. Evidence is increasing all the time. All major scientific schools of thought acknowledge that global warming is occurring, that the enhanced greenhouse effect is a reality, and that its effects are altering the climate.

Excuse 2: Technology will be able to halt climate change

There is an extraordinary degree of wishful thinking about technological possibilities for 'solving' climate change. This blind optimism is shared and encouraged by the political and business community. It is apparent that this form of denial is being used as justification for evading a serious challenge to the status quo.

- Solar panels covering a very small area of the Sahara Desert would supply all the world's needs for electricity.
- The problem of excess carbon dioxide emissions can be dealt with by their sequestration in forests, mines or under the sea.
- Applying the latest technologies could increase four-fold the efficiency with which energy-dependent activities are now carried out, and there is the prospect of future technologies delivering a ten-fold increase.

Response

The scope that technology has for combating climate change is dealt with in detail in Chapter 6. It demonstrates clearly that technology is highly unlikely to be able to reduce carbon dioxide emissions *sufficiently* – certainly not within the timescale available to avoid serious damage to the planet. Major adjustments to our lifestyles will be necessary.

Excuse 3: I blame the government / the Americans / . . .

Criticizing others is very common in the climate change debate. The psycho-logical mechanism at work is 'projection', which occurs when individuals attribute their unwanted feelings and characteristics on to someone or something else so that others can be accused of worse behaviour. Blame is redirected on to a range of targets, be they countries, governments, corporate organizations, previous generations, particularly wayward individuals – or even animals! It is much easier to place the responsibility for combating global warming elsewhere, on the grounds that individuals do not have the resources or the power to bring about the necessary changes.

- Americans are the worst culprits, accounting for a far higher proportion of global emissions than their proportion of the global population.
- The real problem lies in the fact that China's economy is growing at such a pace that within a few years its GDP will exceed that of the US.
- This one is for government . . .
- It is far more important to consider the effects of population growth in the developing countries.
- Flatulence in cattle and sheep is a major source of greenhouse gas emissions.

Response

While all these statements have some validity, the unavoidable fact is that climate change is a global problem. All nations – including the UK, which has been responsible for 15 per cent of total global emissions of carbon dioxide to date and currently for two and a half times global-average emissions per person – must contribute their fair share to its resolution.

Excuse 4: Shooting the messenger

Issue is often taken with the purveyors of 'bad news'. People may attempt to devalue the argument that climate change is serious and that lifestyles need drastic modification, highlighting the behaviour of the 'messengers' either by stigmatizing them as 'scaremongers', fundamentalists, or as hypocrites.

- Global warming is a godsend to alarmists!
- You are an ecological fascist!
- *You* own a car!

Response

Shooting the messenger is always an inadequate and ineffective response. The message is not invalidated by having done so.

Excuse 5: It's not my problem

People can use different excuses to dissociate themselves from the problem of climate change based upon either feigned or real indifference. In addition, many people in society feel disconnected politically, socially and culturally. This enables them to believe that the significance of global warming can be dismissed on the grounds that it is too complex, or that the phenomenal advances in science and technology will mean that future generations can cope with it far more readily than we can.

- We don't have to solve our children's problems – we've got enough of our own.
- I don't care – I'll be dead by the time it happens.
- I don't really understand it.

Response

Climate change is already happening and it is therefore our responsibility to tackle it. This is not a problem that can be left as a legacy to future generations. Individuals should be accountable for the effects that their actions could have, and narrow self-interest or ignorance are obviously inadequate excuses for not being so.

Excuse 6: There's nothing I can do about it!

Helplessness – a manifestation of the psychological response of 'dissociation' – can arise from a number of causes, such as a pessimistic outlook, a dislike of change or, more commonly in this instance, a conviction that individuals are powerless to act and in any case can do little by themselves to avert climate change. With this line of defence, there is no reason even to listen, to take to heart the gravity of the problem, or to accept individual responsibility for contributing to it.

- It won't make any difference if I stop flying or only use my car when no alternative transport is available.
- My life's complicated enough already; it'll involve too much effort making changes.
- It's too late. We can't prevent it happening. We will just have to adapt.
- You may be right but . . .

Response

It is not too late, and there is almost no limit to what we can do about it, both individually and collectively. Moreover, the longer we procrastinate, the greater will be the adverse consequences and the greater the difficulty in reversing the process that has brought them about. See Chapters 7 and 8 for more details.

Excuse 7: How I run my life is my business!

Individuals attempt to justify their inaction by pointing, as the initial and most pertinent reaction, to how any change is likely to impinge upon their own lives in an unacceptable way. Most famously, this argument was used by George Bush Senior in 1992, when as President of the USA he declared that 'the American way of life is not up for negotiation'.

* I have to use my car – there is no other way of commuting. Public transport would be fine if it were more efficient and matched the door-to-door convenience of the car.
* It is all very well to say that it is wrong to fly, but how else can I see my friends and family on the other side of the Atlantic?
* I will do what I like with my life, thank you! It's a free country. Nowadays, you can't enjoy yourself without someone telling you that it's harmful.

Response

The future of the planet is at stake. Our own self-centred and short-term gratification needs to be seen in context. We obviously have no right to make decisions for ourselves without reference to their wider, longer-term societal and environmental consequences.

Excuse 8: There are other more important and urgent problems to tackle

The need for urgent action on climate change is denied by citing other world concerns or crises – such as poverty, conflict, terrorism – which, it is claimed, deserve higher priority. It has been suggested that money could be better spent in improving the circumstances of people in the Third World than complying with the Kyoto Protocol.

* The best way of helping the populations of many developing countries is to provide aid for them so that they can have clean drinking water, modern sanitation and electricity.
* Money would be better spent in promoting trade around the world in order to reduce unemployment.

Response

Acting to reduce the risk of climate change immediately is imperative. The importance of other global problems is not negated by doing so. The world could provide electricity and clean drinking water to all, if it chose to, regardless of the adoption of policies to minimize climate change. Ignoring climate change will not help resolve other global problems: in fact, it is likely to exacerbate many existing very serious problems, particularly poverty, availability of clean water, spread of disease, vulnerability of livelihoods and insecurity of food supply.

Excuse 9: At least I am doing something

Individuals who are concerned about the issue of climate change but are not taking effective personal action console themselves that at least they are doing 'something'. Others attempt to seek reassurance for their unecological lifestyles by citing involvement in other worthy causes.

- I try to use public transport whenever I can.
- I am a member of Friends of the Earth *and* Greenpeace.
- I compost my kitchen waste and recycle my paper – which is more than most people do in my street.

Response

This type of excuse encapsulates the wish to feel good about what we do by projecting our energies into areas of activity which, to some extent, relieve our consciences. They are of course worthwhile in their own right. But care must be taken to ensure that they are seen as relatively marginal. The need for widespread preventive action on climate change is much more demanding, and attention should not be diverted from it.

Excuse 10: We are already making a lot of progress on climate change

People concentrate only on positive evidence of progress and ignore contradictory evidence. This has been described as the 'Mississippi syndrome' – focusing on the small boats struggling upstream rather than the huge mass of water flowing downstream. When it suits, this strategy is widely adopted by politicians.

- More wind turbines than ever before have been commissioned recently.
- Modern aircraft are a lot more efficient than those manufactured years ago.
- The UK economy uses far less energy per unit of GDP than previously.

Response

This is a defence mechanism in which only positive evidence is acknowledged. It induces complacency by allowing people to treat isolated, individual signs of progress as evidence of *sufficient* overall progress. Although the UK has taken a lead on climate change on the international stage, its response to date is still far from adequate. We have only taken the first faltering steps on the path to a low-carbon economy. Public policy is discussed in greater detail in the next chapter.

Conclusions

This chapter has identified the most common excuses for not taking the issue of climate change as seriously as we should, and as the previous chapters have indicated, and has provided responses to challenge the arguments on which the excuses are based. The most common have taken various forms of denial about the gravity of the situation, reflecting a state of disconnectedness, not wanting to know, and a substantial element of wishful thinking that 'the problem will go away'. All this has resulted in a disturbing degree of complacency, enabling people to carry on largely as they have been.

It would seem that attitudes, perceptions and the preferred behaviour of individuals stand in the way of the wide adoption of sustainable lifestyles owing to the difficulties of facing up to unpalatable truths. We do not appear to be prepared to do so, preferring instead to continue down a road that we already know to be leading to ecological catastrophe. However, understanding the psychological processes that lie behind the excuses should not be used as grounds for further delay. Instead, it should motivate us to face up to the situation and move more speedily towards effective action before it is too late.

Part Two: Current Strategies

5. Too Little, Too Late?
Government Policy and Practice

Can we carry on with our lives as usual and leave the government to deal with the problem of climate change? This chapter argues that the answer to this question is an emphatic 'no'. Although the UK government has taken a very important lead on climate change, it does not seem to be prepared to tell us as voters what we do not want to hear: dramatically reducing carbon dioxide emissions cannot be accomplished without significant lifestyle changes as well as making best use of technological advances. Unless we face up to this seemingly unpalatable truth, we will not achieve the necessary progress towards reductions in emissions that is essential to the survival of the planet.

This chapter describes what the government is doing, what it is not doing, and what it must do if we are to move fast enough towards a low-carbon future. It is the government's job to provide collective goods such as national security, clean air, a social safety net and to protect the environment from catastrophe. It will have to provide leadership and commitment to reach this goal, even though the goal can only be met in co-operation with national civil society and business interests and, more widely, through agreement with all the other countries of the world. To do this requires a radical change in policy.

Government policies and responsibilities specifically and society's responsibilities more generally are at the heart of the matter. Looking at government policies cannot mean ignoring the role of individuals. Government is influenced by them, both directly as voters and indirectly through their representatives (such as MPs) and their membership of special interest groups (such as the Consumers' Association and Greenpeace). So, while this chapter is about what the government's actions should be, it is also about what we as citizens and voters should allow it to do and prevent *on our behalf*.

What should be happening

While still enjoying as many of the benefits of energy use as possible, society should, we believe, be aiming to achieve the following steps, in the following order of priority and importance so as to reduce carbon dioxide emissions by 60–80 per cent by 2050:

1. Reduce energy-dependent activity.
2. Ensure energy is used as efficiently as possible.
3. Supply as much energy as possible from low-carbon and renewable sources.

The essential difference between our position and that of government is that the government does not recognize the importance of Step 1. Instead, it is concentrating on Steps 2 and 3. It is opposed to Step 1 because it is worried that it cannot win elections by pursuing this path, whereas the evidence shows that it is imperative that it supports it. Otherwise, the savings from Steps 2 and 3 can always be negated by an increase in energy-related activity (as has been the case to date). Without Step 1, there is no guarantee of the necessary savings. Recognition of this order of priority is, in our view, fundamental. Chapter 7 contains a detailed analysis of how Step 1 could be implemented fairly and effectively for personal energy consumption.

What the government is doing

The UK government has initiated a climate change policy the aim of which is to deliver 'a more sustainable, lower carbon economy'. All sectors of society and industry are expected to contribute. The policy is designed to meet both its Kyoto commitment – a decrease of 12.5 per cent of six greenhouse gases from 1990 levels by 2010 – and its self-imposed target of a decrease of 20 per cent of carbon dioxide emissions during the same period. In addition, it is aiming for 10 per cent of electricity to be obtained from renewable energy sources by 2010, and 20 per cent by 2020. Various methods are being used to achieve this end, including regulation, voluntary agreements with industry, information, education, research and development funding, financial incentives, and other economic instruments. EU policies are also an important and integral part of the government's response to climate change.

The determination of policies on this subject and action on them are the responsibility of many institutions, including central and local government, the Energy Saving Trust, the Environment Agency and the Carbon Trust. Some of them are designed to achieve immediate savings, while others may not deliver significant savings for several years: climate change policy is a complex mix of approaches, delivered by different bodies with varying influences and timescales for delivery.

The government's approach to climate change is made up of many policy initiatives. The most significant for expected energy or carbon savings are briefly described in this chapter by sector – transport, household, and industrial and commercial – and the overall effect is assessed at the end. There are two important questions to bear in mind while reading these summaries:

• Will these initiatives be sufficient to meet existing 2010 reduction targets?
• Are they sufficient to prepare for the imperative medium-term reduction of 60 per cent by 2050?

Transport

The overall aim of the government's transport strategy is to provide a transport system that is 'quicker, safer and more reliable'. While flagging up congestion as a major problem, the government does not have a traffic reduction target. Instead, it expects road traffic to increase by 17 per cent between 2000 and 2010. Despite this, it believes policies currently in place will restrict carbon dioxide emissions in 2010 to around the same levels as they were in 2000. That clearly would represent a farcical degree of progress in delivering on its targeted reduction in carbon dioxide emissions. The two major strands of policy are its Ten Year Plan and an EU agreement with European motor manufacturers on improving the efficiency of new cars in production by 2008.

The Ten Year Plan

This is the government's umbrella strategy for transport that was published in 2000. Many of the measures proposed in it are already in place, ranging from technology support for alternatively fuelled vehicles and financial support for those buying them, to educational and planning initiatives. The plan does not include any policies to restrict traffic, including use of the car, except in a limited number of cases where this is thought necessary to reduce road congestion. The rationale is that road pricing, congestion charging and parking control will restrict the use of cars and lorries. At the same time, however, it is felt necessary to improve the availability and quality of alternatives to public transport. Restricting demand, while seen to be acceptable to manage congestion in special circumstances, is not even discussed in the context of climate change.

Taxes have been used to try and influence driver behaviour, both by limiting mileage through fuel prices and by influencing car purchase. The most celebrated transport tax in the UK was the 'fuel duty escalator' or fuel tax, which was introduced in 1993 to rise annually at above the rate of inflation. However, although it could be seen as a tax aimed at achieving a reduction in carbon dioxide emissions, this benefit was not highlighted when it was introduced. When it was removed from government policy in 1999, and following the fuel tax protests of 2000 which brought much of the country to a halt for several days, government found it impossible to continue with this effective measure. Not only that, but in the aftermath of the protests, the Chancellor pledged in his following budget to cut fuel duty by 4 pence per litre. It was the people and not the government who rejected the fuel tax increase as a policy measure.

However, during autumn 2003, the government once again raised tax on fuel, to little apparent public alarm – but the rise was not presented as an environmental measure. The government did not state what savings it intended to achieve with this new tax, although carbon dioxide emissions from transport in 2010 have been predicted to be 1–2.5 MtC lower than would have been the case without the price escalator in operation from 1993–9.

Other policies on taxing transport remain. New cars are subject to different rates of vehicle excise duty (road tax) depending on their level of emissions, with higher-emitting cars attracting a higher charge. However, in the summer of 2003 there were no petrol cars available (other than two-seaters) in the lowest charge band and only three which qualified for the next lowest. At current rates, petrol cars in the highest charge band pay £160 duty a year, but the lowest-rated cars generally available have to pay £105. It seems unlikely that the difference of £55 per year, equivalent to less than the average weekly household expenditure on transport, will be very effective in guiding decisions on car purchase.

The Ten Year Plan includes targets for promoting public transport usage. A 50 per cent increase in rail use is planned from 2000 to 2010 (just under a quarter of which is expected to come from road users switching to rail), and a 10 per cent increase in bus use. Most of the policies to achieve these increases involve a significant subsidy. This has been provided either in support of better facilities (such as upgrading the West Coast railway line), in introducing urban light rail systems in some provincial cities, and by not requiring public transport passengers to pay the full fares that would be required to cover the capital and running costs of the services.

Car efficiency

The government expects the majority of savings in carbon dioxide emissions from transport by 2010 to be achieved through a voluntary agreement on improving fuel efficiency signed between European car manufacturers and the EU (details of the scheme are explained in Chapter 6). By relying on a largely 'invisible' technical measure to reduce the emissions, the public is not being engaged in the need to make savings. Moreover, it is being lulled into a false sense of expectation that this will be a sufficient step to allow for the continuing growth in car ownership and use in the foreseeable future.

Household energy

For household energy, the policy emphasis is also largely on technological improvement spurred on by regulation or voluntary agreements with industry. Subsidies, funded both by general taxation and through energy bills, are offered to improve home insulation and the efficiency of domestic energy-using equipment.

New housing

The most important influences on new housing are the national Building Regulations, which specify the energy efficiency standard which a property has to meet. These standards (which vary slightly between England and Wales, Scotland and Northern Ireland) were first introduced in the 1960s and are updated and made more rigorous every 5–10 years. With minor exceptions, they control only the fabric of the dwelling, however, and not the energy-using equipment within it.

Although the role of the Building Regulations in raising standards is widely welcomed, there are two common and complementary concerns. The first is that the regulations are not ambitious enough: UK housing still lags decades behind standards achieved in other European countries with comparable climates. The second is that there is poor quality control, with the result that the prescribed standards are not met in practice.

Existing housing

There is no legislation for raising the standards of the existing housing stock, in spite of the fact that much of it is poorly insulated and the equipment used in it is wasteful of energy. However, in order to try to remedy this, the government has introduced several grant and incentive schemes. The two major ones are 'New HEES' (Home Energy Efficiency Scheme) and EEC (Energy Efficiency Commitment). New HEES is government funded and offers a package of free insulation and/or heating improvement measures but only for households in receipt of income or disability-based benefits, and with higher spending limits where the householder is over sixty years of age. It aims to reach almost half a million households. The EEC scheme is the latest incarnation of an obligation on gas and electricity retailers to aid householders in reducing their fuel bills. Savings are achieved most commonly by subsidizing consumer purchase of efficient light bulbs and appliances and the installation of cavity wall insulation. Both schemes have aims that are broader than energy or carbon saving. New HEES and EEC are also intended to reduce fuel poverty (where households would have to spend 10 per cent or more of their income to receive adequate energy services) and avoid risk to health from insufficient heating during the winter.

Other policies

There are numerous other policies targeting energy efficiency and carbon dioxide emissions in the domestic sector, including:

- EU energy labels at the point of sale on fridges and freezers, washing machines, tumble dryers, dishwashers and light bulbs. The UK's Energy Saving Trust is also involved in labelling the best of these and other products, such as boilers, with an 'energy efficiency' recommendation.
- EU-wide minimum efficiency standards on a regulated basis for boilers and fridges and freezers, and on a voluntary agreement basis for televisions, VCRs, digital set-top boxes, washing machines, dishwashers and tumble dryers. The minimum standards for boilers and fridges in particular have resulted in considerable energy and carbon savings, compared with what would have happened without them.

Business and public sectors

The UK government expects the majority of carbon savings in these sectors to come about in response to its economic instruments, particularly the Climate Change Levy and trading in carbon dioxide emissions. Some specific technologies, such as CHP (combined heat and power), are also being encouraged. In addition, regulations are employed, for example, in setting efficiency standards for new buildings.

Climate Change Levy

In 2001, UK businesses became liable for the so-called Climate Change Levy, a new energy tax that adds about 15 per cent to their energy bills. The levy rate is 0.15 pence per kWh for gas, coal and coke, 0.43 pence per kWh for electricity and 0.07 pence per kWh for liquefied petroleum gas (LPG). Many intensive users of energy have entered into Climate Change Levy agreements which reduce their tax until 2013 – if energy reduction targets per unit of output are met by 80 per cent. In order to help businesses meet the targets, there are associated measures such as preferential tax treatment for investment in energy-efficient capital equipment. The Carbon Trust, set up in 2001 and funded by receipts from the levy, also aims to help businesses prepare for a low-carbon future and exploit the commercial opportunities that are on offer.

Carbon trading

The government has introduced a voluntary emissions trading scheme for companies seeking to reduce their carbon dioxide emissions. This is a 'cap and trade' scheme. The 'cap' part means that there is a collective limit on emissions from the companies involved which guarantees savings (compared with what emissions would have been without the scheme). Within the overall cap, each company has an individual target to reach. However, in practice some companies will find it easier to meet and exceed their savings targets than others. Trading between companies reduces the cost of making emissions cuts by allowing those which can make savings cheaply (beyond their target) to sell these reductions to others. The overall amount of savings is guaranteed by the cap while trading allows the savings to be made at the lowest collective cost.

The government has injected £215 million of taxpayers' money to kick-start this scheme. However, research carried out by Environmental Data Services (ENDS – which publishes a respected environment news journal) suggests that many of the commitments on emission reductions made under the scheme are merely those which the companies involved are already obliged to achieve under other legislation or to which they have previously voluntarily committed themselves. The cap has not been set at a level which guarantees further savings. In other words, taxpayers' money has been given to participants to do something they would have done anyway. In this case, it seems that the business community has outwitted government. The environment will not benefit.

Despite this (or perhaps because of it), a general enthusiasm for trading schemes from many businesses and from government remains. At the end of 2002, the EU agreed ground rules for a European-wide carbon trading scheme which may incorporate other greenhouse gases at a later stage. Even if the UK scheme is not successful, trading schemes are likely to feature in emissions reduction policies into the future.

Renewable energy

The key aim of UK governmental policy is to reach a target of obtaining 10 per cent of electricity from renewable sources by 2010. The longer-term development of renewables that are nowhere near to being economic, such as wave power, is also a goal. The main focus is on producing renewable electricity, rather than on other forms of energy, such as heat or transport fuels. Government policy on promoting the renewables consists of five elements:

- An obligation on all electricity suppliers in Great Britain to supply a specific proportion of electricity from eligible renewables. This percentage starts at 3 per cent for 2002/3, rising gradually to 10.4 per cent by 2010/11, and further in subsequent years. This target contributes to a wider EU target of 12 per cent of EU energy, and 22 per cent of electricity from renewable sources by 2010.
- The exemption from the Climate Change Levy of electricity from renewable sources, thereby increasing their price competitiveness.
- An expanded support programme for new and renewable energy, including capital grants and an expanded research and development programme.
- The development of a regional strategic approach to planning and targets for renewables. The difficulty of obtaining planning permission, particularly for onshore wind farms, has presented a major setback for expansion in renewables over the past decade. In response to the 'bottlenecks' in planning, the government is requiring each region of the country to set strategic renewable targets for 2010 which should eventually speed up the progress of the renewables energy programme through the planning system.
- Some small incentive schemes to encourage the use of renewables at the household level, such as subsidies for solar water heating, geothermal energy and solar photovoltaics (see Chapter 6).

While welcoming these new initiatives, many commentators are doubtful that existing policy will be sufficient to ensure the 10 per cent goal in 2010 is met even with the planned expansion of offshore wind power announced in 2003.

What does this all add up to?

In summary, as this section will demonstrate further, current UK policy, although good in parts, is nowhere near bold enough to put the country on a path to a sufficiently low-carbon future. It does not lay the foundation for the radical changes that will be required to meet the government's policy of a 60 per cent reduction in emissions by 2050. The UK's approach to climate change is mostly 'business as usual' with a few technological improvements thrown in. This is not a realistic strategy.

Looked at in detail, UK policy presents a mixed picture. Policy approaches vary considerably between sectors, and many of those expected to make most savings are EU rather than UK in origin. Some of the component elements within them have a good track record and are likely to achieve the targets set, especially the Building Regulations and the minimum efficiency standards for appliances. Others, such as the variable road fund tax for cars, seem timid given the scale of the challenge. At best, they are only symbolic. A previously successful policy, the fuel duty escalator, was reversed when the government came under political pressure. However, climate change is too serious an issue to allow ourselves to be content with what are in effect gestures unlikely to bring about sufficient reduction in carbon dioxide emissions.

Will the 2010 target be met?

The government believes that its climate change policy will be adequate to meet its targets. In 2000, it estimated that policies already in place at that time would ensure a reduction of all greenhouse gas emission of 15 per cent from 1990 by 2010, thereby meeting the Kyoto target. Over the same period, carbon dioxide emissions would fall by 8.5 per cent. However, if all the policies in the climate change programme were put in place, then an estimated reduction of 19 per cent of carbon dioxide emissions by 2010 could be achieved, almost meeting the national target. Table 1 (see below) shows how the policies highlighted earlier are expected to contribute to these savings.

Making 20 per cent savings from 1990 represents 34 MtC, but the savings

Table 1: Expected savings from different government policies

Sector	Policy	Savings by 2010 (MtC)
Transport	Ten Year Plan	1.6
	EU car efficiency voluntary agreement	4.0
Household	Energy efficiency (including New HEES and EEC) and some policies not yet in place	2.8–3.9
Industry	Climate Change Levy, together with agreements with energy-intensive industries	4.5
	Voluntary emissions trading scheme	At least 2.0
Renewables	Delivery of 10 per cent target	2.5

from policies listed in the table add up to less than 18 MtC. Some of the shortfall will come from other minor policies not detailed here, but most of the savings had already been made by 2000 (largely through the switch from coal to gas to generate electricity).

The general consensus is that the UK will meet its Kyoto target. However, many independent commentators believe the government is unlikely to meet its target of a 20 per cent reduction in carbon dioxide emissions. As seen earlier, the ENDS analysis throws into doubt savings from the voluntary emissions trading scheme. The renewables target is widely thought to be unachievable given current policy, and therefore savings will be lower than the 2.5 MtC anticipated. In early 2003, the government's Sustainable Development Commission warned: 'The UK will fall well short of the government's goal for reducing emissions of carbon dioxide, the principal greenhouse gas, unless further measures are taken.' It noted that emissions reductions from the Ten Year Plan are particularly at risk. Its estimate is that, without further policy initiatives, the UK will achieve as little as half the 20 per cent in savings, although the Kyoto target is still likely to be met.

How will this prepare the UK for 60 per cent savings?

The government has yet to outline the policies with which it intends to meet the 2050 target. However, it is still reasonable to ask whether policies to 2010 will help progress towards the 60 per cent reduction target. Many current policies are designed to promote energy efficiency and a switch to lower carbon fuels but there are few, if any, significant policies on reducing energy consumption per se. Policy has not caught up with the harsh reality that efficiency only really matters if it leads to *overall* savings in energy.

Furthermore, there is no sign that the government is preparing the population for the future changes that will be required to prevent serious climate change. For example, in transport most savings are expected from an EU technical efficiency policy for cars independent of consumer choice or behaviour. Although the proposed savings are welcome, unless the public is informed about the need to reduce emissions through lowering mileage and adopting economical driving habits, the improvements will not have a lasting effect. As evidence from the last few decades has shown, the savings will simply be wiped out by the increased distances travelled and resort to higher-performance cars. The government is not being sufficiently bold. There is no evidence that we are firmly on the path to the intended 'more sustainable, lower carbon economy'.

Missing and contradictory policies

What the government is *not* doing is at least as important as what it *is* doing. The most striking and important area that they have not tackled is air travel (the significance of which was discussed in Chapter 3). There are also contradictory policies where priority is given to increasing road transport despite the goal of decreasing emissions from these sources. There are policies with unintended consequences, and areas of policy where one government initiative is cancelling out another.

This section picks out some key examples, and shows how the government's 'climate change strategy' is being undermined by its policies in other areas for which it has responsibility.

Support for higher speeds and longer distances

Government is attempting to meet as much of the demand for travel as possible, where and when it arises, and to widen the opportunities for travel by road, rail and air. In the Ten Year Plan for transport, a spending programme of £180 billion, largely allocated to this end, is proposed. EU leaders too are being recommended to spend over £16 billion of Community funds by 2020 solely on improving road, rail and shipping links in Europe, including a road/rail bridge to Sicily, a motorway from the Baltic to Vienna, and a new railway line from Berlin to Naples and from Paris to Bratislava. 'Further, faster and at lower cost' has been the policy direction of successive governments for decades, and there is no sign of any reversal.

There are three fundamental problems to this approach. The first is that it has no limits: travel time and comfort by road, rail and air can always be improved. It is implied that if the French and Spanish can operate high-speed train services, a motorway network only rarely congested, and airports that have the capacity to cater for the number of passengers wanting to use them, the UK should also be able to do so. This absence of limits makes transport policy a black hole, sucking ever more resources in.

Second, and allied to this, is the fact that the improvements cannot be so designed that they only serve current demand. In practice, they encourage more demand. Indeed, if they do not do so, trouble is foreseen: the number of passengers using Eurostar is about half of what had been forecast before it was 'up and running'. This under-use is seen as a growing cause for concern in political and business circles, and the argument has been made that policy needs to be directed to *encouraging* more travel on it so that it can become financially viable. Third, and most importantly, as a policy promoting energy-intensive activity it undermines the objective of limiting serious damage from climate change.

Support for car-based lifestyles

Although government information messages may suggest people should use their cars less (such as in the 'Are You Doing Your Bit?' campaign), in reality substantial direct and indirect policies and practices continue to be made in supporting car users. The promotion of car-based lifestyles is reflected in the fact that applicants for major planning and land use changes have never been required by central and local government to include the generation of traffic that will stem from such changes or energy-impact statements (or carbon emission statements) based on this. A recommendation to do so was made nearly thirty years ago in a submission to a House of Commons Select Committee inquiry on energy conservation. These statements, for instance in relation to proposals for new hospitals, sports centres, business parks and retail outlets, would cover the projected motorized mileage of staff and visitors using the facilities. That mileage, and the projected carbon dioxide emissions, could then be compared with those from alternative locations or patterns of provision. In this way, the decision would then be based on this crucial issue as to whether the public interest was better served by larger facilities, albeit fewer in number, or more numerous though smaller. It is very likely that the incorporation of such statements would result in a radical shift towards a strategy on more local provision rather than one largely dependent upon car use.

Another prominent example of government policy that continues to promote car use is the rule regarding company cars, which still account for half of all new cars purchased. Their ownership leads to above-average personal travel. Despite years of reform to company car taxation, their purchase is still being encouraged.

A further key example of support for car use can be seen in the roads programme. In 2000, £30 billion was announced for this by the government as part of the Ten Year Plan. This vast sum was intended to pay for nearly 600 kilometres (360 miles) of motorway and trunk road widening and 100 trunk and local by-passes, with more money allocated for improving local roads.

This road-building programme is effectively a return to the discredited philosophy of 'predict and provide', whereby the government forecast the likely increase in road traffic ('predict') and then built roads to accommodate it ('provide'). Experience has shown that providing new roads creates more demand for travel than had previously existed. Widening motorways has been described as 'like digging a ditch in a water-logged field': traffic fills them up in the same way that water fills up the ditch. Building additional road capacity might solve the political problems of government which does not wish to be seen to be presiding over a future of increasingly congested roads and does not wish to restrict car use for fear of being seen as 'anti car'.

Nevertheless, in these circumstances, it is thought that an acceptable matching alternative in the form of public transport must be provided before any limit on the free use of the car is actually introduced. For instance, 'Park and Ride' services from the edge of some towns, in combination with other measures such as restricted parking and pedestrianization, have been adopted as a strategy for preventing too much traffic growth in central urban areas. Even this can be seen to be supporting lifestyles that are largely car-based. As the revenue from these services can rarely cover the costs of their provision, the lifestyles of the car users for whom they cater are facilitated by having their travel by public transport subsidized.

The paradox of public transport

Government support for public transport is generally welcomed by environmentalists. It has many important benefits, providing the only option of longer-distance travel for people without access to a car, improving local air quality and reducing urban congestion. However, it is necessary to take a harder look at the claims of public transport in respect of energy use and carbon dioxide emissions. There are three key, generally unacknowledged problems which undermine its environmental credentials:

- It can be used to subsidize high-energy lifestyles.
- In common with the car, public transport generates additional travel.
- Carbon dioxide emissions per passenger kilometre by public transport on current occupancy levels are not significantly lower than by car (as will be seen in Chapter 8).

Public transport, rail in particular, is used to enable long-distance commuting to work. Analysis of data from the National Travel Survey shows that rail commuters travel twice as far on average as car drivers. The availability of a rail service encourages longer-distance patterns of travel. In addition, it is inequitable as a high proportion of the users are people on higher incomes. It is the already better-off who are able to live higher-energy lifestyles as a result of investment in it. This development continues largely unquestioned. In London, a proposal for a new east–west scheme called 'Crossrail' was given a provisional go-ahead in 2003. A claimed key 'benefit' is that it will permit additional commuting from new housing development for the Thames Gateway and Lea Valley areas (to the east of London) to destinations as far away as Richmond and Heathrow Airport (to the west of London). Extraordinarily, too, rather than building new homes in the south-east of England, it has been proposed that they are built in northern France, with commuting to London facilitated by cheap Eurostar fares.

In both these examples, high-energy lifestyles, specifically to facilitate people living further away from their workplaces, are being seriously contemplated because high use of public transport is seen as acceptable in a way that high use of cars is not. However, improved public transport does not simply offer a form of transport with lower carbon dioxide emissions than the car. Like any improved transport facility, it too generates more travel. This is well illustrated in figures covering passenger travel between London and Paris since the introduction of Eurostar. This fast service has led to a massive increase in the number of people travelling between the two capitals, over and above the number who previously made the journey. Moreover, once the fast rail link through Kent and east London is completed, the number of passengers it carries annually is predicted to rise four-fold over the following twenty years.

It is clear that good-quality public transport encourages additional travel, more spread-out lifestyles, and therefore adds to the problems of climate change. Thus there are grounds for concern about its promotion by government.

Flying blind? Contradictions in air travel

One of the most important 'missing' policies of government concerns air travel. As shown in Chapter 3, air travel is the fastest growth area of energy demand and currently contributes around 8 MtC to the UK's carbon dioxide emissions (and its effect on global warming is three times this). However, the government seems unwilling to challenge in any way the rapidly rising demand for it. It is talked of in glowing terms as being essential for growth and good for the economy. Policy is therefore determined on a 'predict and provide' basis, with the focus of attention on where to locate new airports and how to use existing facilities more intensively. However, there is no doubt that, as the RCEP stated in their report on air travel, 'rapid growth in air transport [is in] fundamental contradiction to the Government's stated goal of sustainable development'.

Not only is air travel not discouraged, it is actually strongly promoted through the provision of subsidies. Currently no VAT is paid on new aircraft, airlines pay no duty or VAT on the fuel used, and airline tickets are free of VAT. The value of these hidden subsidies in the UK are variously estimated at £6–9 billion per year. In Europe, there are examples of regional aid being used to pay airlines to fly to local airports, although this practice is being challenged in the European courts. As a result of this, and the effects of competition in the industry, airline tickets are almost half the cost in real terms of what they were ten years ago. As with rail travel, the subsidy particularly benefits wealthier people – 5 per cent of the population account for over 40 per cent of air miles.

Further subsidies are given to support the air industry. For example, in 2001 the government gave a loan of £530 million to fund BAE Systems' (formerly British Aerospace) involvement in the Western European countries' production of a new super-jumbo Airbus. Thus the low price of air travel and its increasingly wide take-up have been achieved by direct and indirect public subsidies with an outcome which is severely damaging to the environment. This is in complete contradiction to the government's avowed support for the 'polluter pays' principle.

Subsidizing energy use

Energy use is subsidized in many ways. The subsidies can be direct and indirect in the form of grants and loans offered at preferential rates. Norman Myers and Crispin Tickell have estimated that 'every year the government gives some £6 to £8 in fossil-fuel subsidy for every £1 to support clean and renewable energy'. Nuclear energy has been in receipt of considerable public funds over the last few decades, with the prospect of that continuing well into the future. The costs of decommissioning its power stations are not covered by the industry and it is exempt from liability for major accidents. Subsidy and tax relief are also provided for the oil industry to encourage continued exploration in the North Sea. Much of the renewable energy industry would be unable to survive if it did not receive substantial subsidies, paid for through electricity bills.

Subsidizing practices are particularly evident in favouring public transport. Strong support has been given in the UK to the nationwide rail modernization programme of over £60 billion for the coming decade, much of it from public funds and at a rate roughly double that spent on it in the 1990s. Indeed, rail fares would have to more than double to eliminate the need for subsidy for the rail industry from the taxpayer. All of these subsidies encourage energy use by promoting travel.

Perhaps more important than the direct subsidies paid to make energy cheaper at the point of use is the fact that consumers do not pay its full costs. Research shows that energy use creates 'external' costs – that is, costs to the environment, not to mention costs to society – which are not captured in the price. In the case of transport, these arise from traffic collisions, damage to local air quality, noise and, of course, the release of greenhouse gases.

Estimates of the true costs of transport vary, but one study by the European Environment Agency suggested that transport users were paying only 30–40 per cent of them. (There are many moral and practical concerns about determining monetary values for the future damage of climate change, making such estimates questionable.) By not requiring these environmental costs to be borne by the user, and by combining this with generous subsidies for public transport – which, as will be seen in Chapter 8, is only marginally less energy-intensive than car use – the outcome is that *motorized* travel is effectively promoted just at a time when local, non-motorized patterns of activity should be encouraged.

Telling us what we need to know

Without the right information, consumers cannot make choices to save energy, either through purchasing equipment that uses less energy or is more efficient or by behavioural changes to reduce wastage. But the availability of this information is generally limited. For products, the biggest advance has been the adoption in recent years of the European energy labelling directive for some appliances and light bulbs. However, there are still many significant gaps in the provision of helpful consumer information, notably for homes, whose installed space and water heating systems account for around 80 per cent of domestic energy:

- Energy information is supposed to be provided with new homes. However, in 2003 researchers found that only 2 per cent of housebuilders provided such information for owners of new homes, although homebuyers are legally entitled to this.
- Energy information is not automatically provided when buying an old house. Despite decades of debate, it seems that energy conservation is not reported by estate agents to be a major concern of prospective purchasers.

In an era of supposed consumer choice, the data needed to make informed choices that take the environment into account are often missing or incomplete. Labelled products do not necessarily provide the most relevant information. New cars now have to display energy-use labels at the point of sale. However, even the most fuel-efficient cars with the lowest levels of carbon dioxide emissions are churning out more emissions than they were tested for because fuel consumption measurements on new cars cover neither the weight nor the operation of air-conditioning equipment where this is included in the vehicle.

Similarly, the information offered to householders on their energy bills is hopelessly inadequate. Despite the privatization and liberalization of the UK energy market, bills are often estimated rather than based on actual meter readings, they are only received months after consumption has taken place, and offer no specific information to the customer about the role of each component of overall consumption and how it could be reduced. There have been few innovations in the provision of information and advice to house-holders. This is a valuable but missed opportunity. Research has shown that good feedback or information enabling comparison to be made between subsequent billing periods, or against the same period of the previous year, or with the national average, can *by itself* lead to significant reductions in energy demand which are maintained over time. No further interventions are required. The energy industry has shown little interest in providing this valuable information. Government could require it to do so.

Bias against renewable energy and combined heat and power

As noted earlier, the government has policies in place in support of the renewable energy industries, and these are responsible for much of their success. However, there are also policies that are working against their devel-opment. The key example is the reformed electricity trading market, known as NETA (the New Electricity Trading Arrangements). This was designed to lower wholesale electricity prices, but a side-effect is that renewable energy sources have been disadvantaged due to the trading rules adopted. NETA also harms the development of CHP – combined heat and power schemes in which the heat is retained for use rather than thrown away, as in conventional electricity power stations, and in which more 'useful' energy is thus produced for the same level of carbon dioxide emissions. The government has chosen not to change NETA, and so one strand of government policy is opposing another. While government policy is pulling in two directions at once, it cannot expect to be effective. Revising policies that are likely to lead to increasing carbon dioxide emissions may be at least as effective as devising new policies to reduce them.

Why is the government sabotaging its own carbon-saving efforts?

The government is aware of at least some of the contradictions in its own policy. It has openly admitted that increases in air travel by 2010 could negate half the carbon savings it expects to make from its climate change policy by that date. Yet somehow this recognition has not spurred it into action. This may be because it can still reach its Kyoto target if aircraft emissions continue to be excluded from the calculations. If that is the reason, it would mean that, despite its rhetoric, the government has not really taken to heart the seriousness of the implications of climate change and the major contribution that air travel is making to this.

Another theory explaining why the government is ignoring the many contradictions in its policies is that to acknowledge them would entail questioning the role of economic growth and consumer choice, which it is not prepared to do. Nor is it alone in this: there is plenty of evidence to show that individuals, too, do not want to face up to the consequences of their energy-dependent lifestyles (see Chapter 4).

The hidden issue: economic growth and energy

In the previous section, many of the areas in which government policy is contradictory or missing were shown to be ones that promote increased energy-based activity. As identified earlier in the chapter, this is the key difference between the authors' views and those of the government. The government believes that an increase in energy-based activity does not necessarily contradict its policy on achieving major carbon savings, and that reductions in carbon dioxide emissions will be largely provided through technological change. This is clearly illustrated in a 2003 speech by Tony Blair where, talking about carbon dioxide emissions, he stated: 'If we harness new technology, the evidence is mounting that we can achieve a target of 60 per cent – and at reasonable cost.' Chapter 6 reviews the evidence for this view and finds it wanting.

Although successive governments have had energy efficiency policies for many years, and have pledged action on climate change, their more prominent objective continues to be to promote growth in the economy. This is seen as the only way of providing the public with both improvements in quality of life and an extension of choice. Any downturn in the economy is seen as a serious cause for concern because the wealth created from economic growth is used to provide essential services in health, housing, education and other sectors. However, the essential services provided by the planet, such as a habitable environment free of serious damage from climate change, are taken for granted and thereby put in peril.

Squaring the circle?

There is a widespread belief that economic growth and environmental improvement can be pursued simultaneously and, further, that improved environmental conditions are only possible in an economy which is burgeoning. This belief goes by many different names and is reflected in a variety of terms, including *decoupling* (separating economic growth from energy use); the *triple bottom line* (improving economic, social and environmental performance simultaneously); *win-win solutions* (beneficial both economically and environmentally); *sustainable development*; *ecological modernization*; and *Factor 4* and *Factor 10* (improving the efficiency of fuel use four- and ten-fold). These philosophies rely largely on a highly optimistic view of what and how speedily technological progress can deliver, in combination with the concept of 'weightless' forms of economic activity (which place no burden on the environment).

Proponents of the potential for decoupling energy and economic growth point to the fact that, over time, developed economies have tended to use less energy for each unit of GDP produced. Government figures show that between 1990 and 2000 the economy expanded by 26 per cent while final energy demand grew by only 8 per cent. However, all that this actually means is that energy use is growing at a slower rate than economic growth. It does not offer any hope that energy use will actually start falling as the economy grows as part of some sort of natural progression. Moreover, there is a significant degree of lazy thinking in the idea that there are no limits on the extent of decoupling. Emissions largely reflect the rise in economic activity and are too closely linked to allow for complacency on this front.

Both technological over-optimism and hope for a 'dematerialized' economy reflect wishful thinking rather than the attitudes of responsible government. As Chapter 6 argues, there is no evidence to indicate that developments in technology alone could achieve 60–80 per cent of carbon savings by 2050. Most material prosperity is grounded in the use of finite resources: as we have got richer, we have used more energy and acquired more possessions, rather than the reverse. The 'new' economy based on IT and the internet has turned out to be similar to the 'old' economy. A recent Europe-wide report showed that information and communications technology offers few opportunities to 'dematerialize' Europe's economy, and could actually increase the use of resources. As the government's Sustainable Development Commission reported recently: 'The overwhelming consensus amongst academics is that resource productivity will not, on its own, deliver the desired reconciliation between the pursuit of economic growth and the imperative of learning to live within the Earth's biophysical constraints and carrying capacities.'

Sacrificing the planet to save the economy?

For those who are not convinced that high levels of economic growth can be made compatible with reducing greenhouse gas emissions, there are essentially two options:

- Continue to prioritize economic growth, and adapt to the consequences of climate change.
- Prioritize policies aimed at a major reduction of greenhouse gases to minimize the risk of serious climate change, and adapt to the consequent new economic circumstances.

Supporters of the first position either directly argue or indirectly imply that rather than preventing further climate change, we should adapt to it. Adaptation measures would include constructing higher sea walls, building more reservoirs, constructing buildings to withstand an increased intensity and frequency of storms, and providing more help to developing countries. Some economists in this camp have suggested that human wellbeing would be better served by high levels of economic growth, the benefits of which can be used to pay for adaptation to climate change, rather than by having lower economic growth and a less damaged climate. They argue that the cost of preventing climate damage is greater than the cost of the damage itself. Given the climate risks outlined in Chapter 2, this is clearly nonsense, and dangerous nonsense at that. As a Danish energy expert has said, pointing out the limitations of economic analysis: 'It may not be cost-effective to save the planet, but we should do it anyway.'

The extreme version of this position is heard less often now (at least in Europe) as the risks of climate change have become clearer. However, there is still an almost universal belief that economic growth is the most important goal for society and that preventative measures for climate change cannot be allowed to compromise its pursuit. The outcome of this belief – often unexamined – may not be very different from the one that the more extreme economists have proposed.

Instead of sacrificing the planet to save the economy, the answer must lie both in taking effective action to reduce emissions and reconsidering the role of economic growth. As the Sustainable Development Commission has suggested, what we need to do is to look at ways of decoupling improvement in people's *quality of life* from increases in the use of resources. Economic growth is a tool for improving quality of life and wellbeing, but it is often forgotten that it was never meant to be an end in itself. Beyond a certain level, more money does not buy greater happiness: international surveys of happiness show that once a country has income levels in excess of about £10,000 per person, its level of happiness is not related to the level of personal income. In other words, richer societies are not necessarily happier ones. In the UK, research shows that despite income levels rising by 80 per cent since 1970, self-reported life satisfaction has not increased at all over the period. We need to reconsider what it is that we are gaining from increased economic growth, and safeguard the essential gains from it only in so far as it does not prejudice the future environment.

This book does not have all the answers to what is the right balance to be struck between the two objectives of achieving economic growth and ensuring a stable climate. However, it is very clear that any economic approach that endangers the future of the planet, as our current model is doing, is unacceptable, no matter how much wealth is generated.

Conclusions

To summarize the main themes of this chapter:

- Government has committed the UK to reducing carbon dioxide emissions in the long term with a target of a 60 per cent reduction by 2050.
- Present policy is in place to meet shorter-term reduction targets to 2010. The Kyoto target is likely to be met, but independent observers doubt the government is on target to meet its promise of a 20 per cent reduction in domestic carbon dioxide emissions.
- Current government policies are not leading towards the more fundamental changes that will be required for a sustainable, low-carbon economy. More seriously, many of its policies (and subsidies) are encouraging the development of a higher-energy society, precisely the opposite of what is needed. Air travel policy is a perfect illustration of this failure.
- The government is unrealistically optimistic about the role of technology in providing reductions in energy use and carbon dioxide emissions.
- There is no evidence that reductions can be achieved solely by win-win strategies, where a policy on reducing emissions can be made compatible with conventional economic growth.
- Economic growth is simply a means of achieving human wellbeing. So is reducing carbon dioxide emissions. There will have to be a rebalancing between the economy and the climate so that the climate, and our future, is not sacrificed to save the economy.

This chapter has highlighted the many incompatibilities and inconsistencies of UK government policy. (That is not to say that other governments have demonstrated that they are free of similar policy failures.) We can no longer allow ourselves the luxury of this sort of double-think. We have to face the reality of climate change, and acknowledge the gap between the aspirations and delivery of government policy as well as its fatal internal contradictions. This will mean making hard choices and overturning long-held views. But we have no choice. There is no time left for believing that technology will get us off the hook, as we shall see in the next chapter. Nor should we be indulging in environmental tokenism. Radical changes in government priorities and policy are required to prevent serious climate change, and we as voters have to press for and accept the changes that this entails in order to succeed.

6. Wishful Thinking
The Role of Technology

There is widespread hope that technological advance will be the key to resolving climate change. This chapter looks at the most promising contributions that it can make to reducing carbon dioxide emissions. The fundamental question is whether they can allow us to continue with our current energy-intensive lifestyles and increasing dependence on energy, or whether they must be combined, or even subservient to, policies on reducing energy-based activities in order to achieve the necessary savings.

To reduce carbon dioxide emissions by 60–80 per cent by 2050, technological options either have to enable lower use of fossil fuels, through efficiency or exploiting non-fossil fuels, or capture and store the carbon dioxide emitted from fossil-fuel use. There are a number of approaches which can do one or other of these. The options assessed in the chapter are as follows:

- Renewable energy.
- Nuclear energy.
- Energy efficiency.
- Alternative vehicle technologies (such as electric and fuel cell cars).
- The hydrogen economy.
- Carbon sequestration (the removal of carbon dioxide from the atmosphere).

All of them appear to have something to offer in helping reduce carbon dioxide emissions without us having to change our lifestyles. However, they also all have limitations, particularly when set against the huge carbon savings which must be made. Each option is examined to assess its current and possible future savings potential.

At the outset, however, it is important that technology is not blithely seen as the answer to climate change. Determining the role it can play must be realistic. We must be honest about what it can do for us. After all, there is good reason to be sceptical even before looking at the detail. It is technological advance which has enabled us to discover, exploit and find increasing uses for fossil fuels. Less technologically advanced societies generally use less, not more, energy and emit less carbon dioxide. So to what extent can we use technology to move towards a low-energy and low-carbon future?

Renewable energy

Renewable energy makes use of natural energy flows and sources in the environment, such as wind, waves, running water, sunshine and biomass (plant matter), which, since they are continuously replenished, will never run out. Renewable energy emits zero or low levels of carbon dioxide and so seems to be an ideal replacement for fossil fuel if acceptable ways can be found to harness enough of it.

This section looks at current forms of renewable energy in the UK and the technologies with the best prospects, in order to see how much fossil-fuel energy could be replaced and how quickly. The major focus is on the forms which could contribute most in the next twenty years: wind power, biomass and waste.

Renewable energy sources in the UK

Renewable energy, as with other energy sources, is used in three fundamental ways: as fuel for heating, fuel for motor vehicles and to generate electric power. In the UK, most renewable energy is used to generate electricity (more than three-quarters at present). Almost all of the remainder is used to produce heat, with a very small amount as a transport fuel ('biodiesel'). The concentration on using renewables for electricity generation has increased over recent years and is expected to continue. This is in part because renewable energy, which tends to be relatively costly to produce, can compete better against fossil sources in generating electricity, the more highly priced form of energy.

In total, renewable energy in the UK contributed just 1.4 per cent of primary energy demand in 2002. This compares with 1.2 per cent in 2000. However, there is considerable scope for increasing its share more speedily than in recent years.

Electricity

In 2002, renewables (including waste) were used to generate only 3 per cent of all electricity used in the UK. Large-scale hydropower was the dominant source, producing 40 per cent of all renewable electricity. Over the past ten years, its contribution has remained fairly static with any variation largely attributable to differences in rainfall. There has been strong growth from most other renewables, however, as can be seen in Table 2 (see below). The most spectacular rate of increase since 1992 has been in wind power: its energy output has risen almost fifty-fold, with the result that ten years later it supplies 11 per cent of renewable electricity. However, in terms of electricity generated, landfill gas has increased its contribution most over the period, reaching 23 per cent of the total in 2002. Almost half of all renewable electricity is produced from biofuels, such as wood, and waste.

Heat

Generation of heat from renewable resources reached a high point in 1996 but since then has been in decline. This is mainly due to a drop in the combustion of wood for industrial purposes, a response to more demanding legislation on air-quality emissions standards. Nearly all renewable heat is generated from biomass and waste, with only 2 per cent from active solar heating. The most important source is wood combustion in both industrial (38 per cent) and domestic (29 per cent) settings, followed by, in order of importance, straw combustion, sewage sludge digestion and waste combustion.

How can the contribution of renewable energy be increased in the UK?

The government's target is that renewable energy should supply 10 per cent of electricity by 2010. It was only 3 per cent in 2002. In the next two decades, biomass and wind turbines are very likely to be the two main areas of growth. Further into the future, other technologies such as active solar heating, solar photovoltaics and wave or tidal power may be able to make a significant contribution. Hydropower is unlikely to experience significant growth. There is only one community-scale British geothermal scheme, which taps into underground heat, and little likelihood of expansion.

Wind power

A wind turbine is probably the image that comes most readily to mind when people think of renewable energy. Using the wind for power has a long tradition in many countries – in the eighteenth century, Britain had around

Table 2: Electricity generated from renewable sources and waste, UK, 1992 and 2002 (GWh)

Source		1992	2002	2002 % total
Wind		33	1,256	11.0
Solar photovoltaics		0	3	0.0
Hydro	Large-scale	5,282	4,584	40.1
	Small-scale	149	204	1.8
Biofuels and waste	Landfill gas	377	2,679	23.4
	Sewage sludge digestion	328	397	3.5
	Municipal solid waste combustion	177	958	8.4
	Other biofuels and biodegradable waste	52	870	7.6
	Non-biodegradable waste	104	494	4.3
	Total biofuels and waste	1,038	5,398	47.2
Total renewables		6,502	11,445	100.0
Total of all electricity generated (excluding imports)		319,961	385,742	

10,000 windmills. Since the early 1990s, growth in wind-power production has been rapid, and it is an oft-quoted fact that the UK has the greatest potential for wind-energy generation in Europe. However, this potential cannot be fully exploited. First, the wind does not blow all the time, nor does it blow at the right speed for the turbines (they cannot operate in very low or very high winds). This leads to intermittent generation even with wind turbines distributed across different locations. The generally accepted view is that 20 per cent of national electricity is about the maximum that wind could contribute without risking serious supply shortages at certain times. However, in the longer term, developments in energy storage may reduce the importance of this constraint.

Second, power generation varies with the 'cube' of the wind speed (wind speed multiplied by itself three times), making high average wind speeds critical to success. In practice, this often requires the location of turbines in upland, often remote and visually and environmentally sensitive landscapes, which has been problematic. Nevertheless, wind power is the renewable

source with fewest technical, economic and environmental obstacles to overcome before it can make an appreciable contribution to national supplies.

Onshore wind power

Early in 2003, there were just over one thousand onshore wind turbines in the
UK. A typical, modern onshore wind turbine of 1.3 MW will produce enough
electricity annually for 800–1,000 homes. Therefore supplying, for example,
one-third of UK households with wind-powered electricity (equivalent to
10 per cent of total national electricity) would require almost 10,000 additional
wind turbines, ten times more than the current number.

Wind power is already comparable in cost with coal-fired electricity generation, but it still tends to be more expensive than the cheapest option, which is
high-efficiency gas generation. The European Wind Energy Association claims
that, in the twenty years up to 2002, wind power costs had fallen by 80 per
cent, and they are expected to keep falling into the future.

Offshore wind power

At present there is considerable interest in the potential for offshore wind
turbines. As costs are higher than for onshore ones, the key factor driving
interest in offshore wind power is the avoidance of visual intrusion, and
hence an easing of difficulties in obtaining planning permission for them. UK
government data suggest that offshore wind farms around the coast could
provide around one-third of our annual electricity needs. Other assessments
point to offshore wind providing more than the UK's current total demand for
electricity.

At present, little offshore wind power is generated anywhere in the world.
The UK has built 4 MW of wind power at Blyth in Northumberland, with two
turbines close to the coast. The first sizeable development, which should be
completed by the end of 2004, will be of thirty 2 MW turbines at Scroby Sands,
east of Great Yarmouth in Suffolk. More projects are planned, and in summer
2003 the government opened up further areas of sea for wind power. If
developers take up all the available opportunities, offshore wind power from
these areas could provide between 3.5 and 5.5 per cent of the UK's electricity
requirement by 2010. However, the construction of these turbines cannot be
guaranteed, partly because potential investors are wary by virtue of the fact
that they cannot be assured of government policy on the price of electricity
throughout the active lifetime of the turbines. Recent history shows that the
price of electricity has gone down in the face of competition.

Offshore wind power currently delivers electricity at about twice the price
of good onshore farms. Installing wind turbines offshore entails additional
costs – for installation, higher maintenance, making the turbines robust

enough to survive the marine environment, and providing undersea cables back to the onshore grid. However, some of these costs can be offset by the advantages of higher wind speeds and more stable wind regimes, leading to higher electricity output. There is still scope for technological improvement, and costs are widely expected to fall, although they are unlikely to fall below those of onshore wind power.

Waste and biomass
Although less celebrated than wind power, waste and biomass currently produce almost half of the UK's renewable electricity (more than four times as much as wind power) and nearly all its renewable heat. The supply of both waste and biomass is much more predictable than wind, and both sources provide energy on demand. In this respect, they resemble the fossil fuels on which the current energy infrastructure is heavily based. However, the potential for growth is not thought to be as great as for wind power, due to limits on the supply of waste and on the availability of land for growing energy crops such as willow and fast-growing grasses.

Waste
Much of the debate about the prospects of waste as a renewable source of energy stems from concerns about the environmental and health impacts of some waste management options, particularly incineration. Landfill gas, approximately half methane and half carbon dioxide, is produced from decomposing organic material (animal and vegetable matter, paper, wood), which is a renewable resource. The majority of municipal waste that generates energy when it is incinerated has the same renewable organic origins, but in addition the plastic, oil-derived component of waste also provides energy (this is counted under 'non-biodegradable waste' in Table 2 along with electricity produced from waste tyres). Thus all landfill gas energy and most energy from municipal incinerators is renewable in origin, and will save carbon compared with energy derived from fossil fuels.

Of all the renewable sources of energy, landfill gas has been the greatest success of the 1990s. It accounted for around half of the total increase in electricity generated by renewable sources between 1992 and 2002, significantly outperforming wind or any other source of energy. Landfill gas offers the cheapest means of generating electricity from a renewable source – the key factor in its success. This is due in part to the need to collect gas from landfill sites, on safety and environmental grounds, whether or not it is used to create energy. It can be used to produce electricity, heat or a combination of the two. In the next two decades, landfill gas utilization is expected to increase but, in the longer term, perhaps after 2025, it will decrease. This is because of the expected reduction in the amount of organic waste going to landfill sites from which the gas is taken, primarily as an outcome of European legislation.

The recovery of energy from waste incineration is expected to increase in the UK, particularly in response to the requirement to reduce waste going to landfill sites, but there is still uncertainty about how many new incinerators will be built. In addition to municipal waste incinerators, energy is also recovered by incinerating 'specialist' waste, such as clinical waste, tyres (which are not from a renewable source) and chicken manure. The prospects for further energy recovery from this source largely depend on waste management economics and policy rather than on energy policy.

In general, production of energy from waste, whether via landfill gas or incineration, should increase with more waste generation and is expected to continue to do so. Government projections suggest municipal solid waste (a small percentage of total waste) will grow at 3 per cent per year. There is also likely to be further exploitation of sewage gas and waste streams from agriculture and forestry. Therefore waste will continue to be an important and growing contributor to the UK's renewable energy supply. However, increasing quantities of waste should not be welcomed. Although the UK should maximize the amount of energy recovered from it, consistent with recycling and re-use targets, waste reduction is the single most important goal in environmental waste management. Waste represents wasted energy (the energy originally used in the production, transport and retailing of the now waste products) as well as wasted material resources. A low-carbon society would also be a low-waste society. A reduction in the waste available from which to create energy would be more beneficial than increasing energy production from increasing amounts of waste. While this book focuses on energy, changing the way we manage material resources is also a crucial component of a low-carbon future.

Biomass

'Biomass' refers to any plant material, and biomass energy or 'bioenergy' refers to plant material that is used for fuel. This may be specially grown as an energy crop or – in the vast majority of cases – is derived from a waste stream in an industry unrelated to energy production: forestry waste, sawmill wastes, straw and chicken litter. The classic energy crop is trees, but other plants, such as grasses, are also being investigated. Plants absorb carbon dioxide when growing and, when they are combusted, this carbon dioxide is re-released into the atmosphere. Hence there is no net carbon dioxide emission over the lifecycle of the plants. However, use of artificial fertilizers, transport to the point of use and processing will usually cause some associated emissions of carbon dioxide, so these inputs must be kept to a minimum.

Most biomass is currently used to produce heat and relatively little for electricity generation (a small fraction of the 'other biofuels and biodegradable waste' shown in Table 2). By far the largest contribution comes from wood burning for industry and in homes – the 'low-tech' end of the sector. It is estimated that around 3 per cent of UK wood production is used as fuel. UK forests currently meet only 15 per cent of our wood and paper needs, the remainder being imported, so that any potential for increasing wood production for fuel from UK forests would have to compete with demand for paper pulp, wood board and furniture. There is also a small number of energy crop schemes, primarily involving coppiced willow, as well as a grass, *Miscanthus*. Hopes are high for the further exploitation of energy crops to generate electricity but, at present, the technology associated with developing systems for growing and harvesting is in the prototype stage. It is therefore early days to be looking to biomass for electricity generation, and it remains to be seen whether it will ultimately prove to be a more fruitful approach than providing heat from biomass.

Biomass could also be used to produce transport fuels with low carbon dioxide emissions. These biofuels have the advantage that, after being processed from biomass, they produce only about one-third to half the emissions of petrol. They come in two main types: as 'biodiesel', which can be used as a direct substitute for fossil-fuel diesel, and as alcohol, which can be blended with petrol in various formulations. Biodiesel is an oil which is extracted from various crops, most commonly rapeseed or sunflowers, and then treated. Alcohol (methanol or ethanol) is produced from the fermentation of sugars and starches from various crops, for example sugar beet. However, the potential for biomass production is severely limited by competition for arable land from food production and other land uses. Even in the unlikely event of a quarter of UK agricultural land being used for producing biofuel, it would displace only 12–30 per cent of current demand for road transport fuels. Given that this quarter of the agricultural land would represent two-thirds of all arable land, it is highly unlikely that so much biofuel would be produced. Thus the potential for replacing transport fuels with renewable alternatives grown in the UK is very limited. It is unlikely to reach even 1 or 2 per cent of current demand in the foreseeable future.

Other renewable technologies

The three other renewable technologies that currently seem most likely to make a substantial contribution to UK renewables in the next few decades are wave power, solar photovoltaics and solar water heating.

Wave power

Wave power intensity around the coasts of the UK is among the highest in the world. However, devices to make use of wave power are still in the prototype and testing stage. Although some very optimistic estimates have been made of its potential contribution, it seems too early in the development of the technology to place much credence on these. Certainly, wave power is not expected to be a significant source of UK electricity within the next twenty years.

Solar photovoltaics

Solar photovoltaics (PV) turn energy from the sun directly into electricity. There has been considerable research and development of this technology and efficiency has risen sharply and costs have fallen. However, it is still a very expensive way to generate electricity under UK conditions (the same does not apply in developing countries in areas which lie beyond the reach of the electricity network, where solar PV may be the cheapest option). A solar PV system installed on a roof in the UK typically costs £10,000–15,000, but will supply less electricity than that used by the average household and does not pay back its costs within its own lifetime. Costs are expected to continue to fall, but PV is unlikely to make a very significant energy contribution within ten to twenty years due to its cost. Its future contribution will depend greatly on how much its costs can be reduced. At the moment, it is too soon to know whether PV will make the breakthrough to being a significant electricity supplier, or if it will remain a niche product used in special situations where other electricity sources are unavailable.

Solar water heating

Solar water heating for domestic properties and swimming pools is estimated to produce 2 per cent of the UK's renewable heat. This is quite a mature technology with few expectations of significant decreases in price. On a favourably oriented roof (facing due south with a slope of about 45°), it can provide up to half a household's annual hot-water needs. However, it is generally competing in cost terms against gas, which is very cheap, and so it is not usually a good economic investment for the individual householder. Costs vary, but around £2,000–4,000 is typical for an installed system. Of these three options, solar water heating could most quickly and cheaply make a useful contribution to providing energy in the UK, but it would need more government support in the form of generous grants to do so.

The future of renewable energy

The difference between optimistic and pessimistic views on renewable energy is considerable: some claim that all our electricity needs can be provided by offshore wind turbines, whereas others doubt renewables will ever supply a significant proportion of our energy supply. Left purely to the market, relatively little additional renewable energy would be employed for generation, simply because it is generally more expensive than the gas power station alternative. This would not be the case if the full environmental costs of fossil fuels (including oil spills, local air pollution and climate change) were charged to their users, but at present they are not. The future of renewable energy in the UK depends very much on government policy (discussed in Chapter 5) more than on other factors, such as improvements in renewable technology or variation in fossil-fuel prices.

Without a truly radical reordering of governmental and social priorities, the maximum renewables are likely to contribute by 2020 is 20 per cent of electricity supply, and achieving this is not going to be easy. One-fifth of the electricity supply would represent around 10 per cent of our primary energy needs. In addition, in the domestic sector, with enough investment, solar hot-water heating could supply up to a quarter of hot-water needs. There is likely to be only a small further contribution to heat or transport fuels from biomass and biofuels by 2020 – even supplying 1 or 2 per cent of the energy used in these sectors would be ambitious. By 2050, renewable sources should be contributing more to the UK's energy supply on economic grounds alone, however, as costs should have fallen as the volume of renewables rises.

Care must be taken with statistics based on the proportion of energy that can be provided from different sources. The easiest way of increasing the *percentage* of electricity from renewables – and at no cost – would be to reduce electricity consumption! The lower the total electricity used, the more likely renewables can provide a significant proportion. In an ideal world, with better storage systems and much lower levels of electricity demand through improved efficiency of production and changed lifestyles, renewables might be able to supply the majority of our electricity needs. Production of renewable heat from biomass and biofuels for transport, on the other hand, is always going to be very constrained due to shortage of land on which to grow energy crops in the UK.

Achieving increases in renewable energy is a goal we should aim for, but in the wider context of the huge reductions in carbon dioxide emissions required and over such a short period, it is likely to represent only a fraction of what is needed.

Nuclear power

Nuclear power offers a source of electricity with near-zero emissions of carbon dioxide and on a scale which could replace much of the electricity presently generated by fossil fuels in industrialized countries. Over thirty countries make some use of nuclear energy and, of these, seventeen rely on it for more than one-quarter of their electricity, with France and Lithuania having the highest dependence, at around 80 per cent. In the UK, nuclear power provides over a quarter of our electricity (roughly one hundred times more than the proportion supplied by wind power). However, its use involves environmental, health and safety risks which many think are unacceptable. In addition, the financial cost of building and operating nuclear power plants is higher than most alternatives, even before considering the likely vast costs of decommissioning old stations at the end of their lives and the long-term management of nuclear waste.

Despite its carbon-free credentials, nuclear power is not widely seen as a strong contender for supplying future UK energy needs. Currently, global construction of new power stations is largely confined to some nations in the Far East. In some European countries there are moves to close nuclear energy plants before the end of their lives. The last nuclear reactor to be built in the UK was Sizewell B, which came into full operation in 1996. At present, no new ones are planned and, owing to their limited life, if none are built, nuclear generation capacity will decrease to zero by around 2035. As its contribution declines, carbon dioxide emissions from electricity generation will increase, unless all current nuclear capacity is replaced by renewable sources. Recent government statements have not given a lead on the future of nuclear power, stating only that it is important to 'keep the nuclear option open'.

What are the main advantages and disadvantages of nuclear power and what future role could it make towards attempting to deliver 'zero carbon' electricity?

Benefits of nuclear power

The primary benefit of nuclear power is that it can offer electricity with near-zero emissions on a large scale. It can also continue to do this over a long time as considerable supplies of uranium exist in the ground – potentially over 250 years' worth at current usage rates. Moreover, there is the possibility of extracting additional uranium from sea water, or of generating increased quantities of plutonium (another nuclear fuel) via specialized nuclear reactors.

Disadvantages of nuclear power

The primary disadvantages of nuclear power arise from the variety of risks associated with it. The risks due to routine releases from power stations are hotly debated, with a long-running campaign in the Republic of Ireland calling for the cessation of nuclear discharges from Sellafield into the Irish Sea. There is no disagreement with the fact that accidents at nuclear plants can be disastrous, as shown by Chernobyl. Although stringent UK and international regulations mean that the risk of accidents happening at a Chernobyl level is very low, the potential consequences are almost too disturbing to contemplate. The risk of terrorist attack seems much more pertinent since '9/11', although it has long been part of the discussion on nuclear power. Much attention now focuses on the threat of an attack from the air. It has been calculated that if an aeroplane strike released just half of the nuclear waste stored in one of the Sellafield buildings, the release of radioactive caesium 137 would be forty-four times as great as in the Chernobyl catastrophe.

However, although these are all issues of serious concern, the one which has been identified as the biggest barrier to the expansion of nuclear power is the question of nuclear waste. Nuclear waste requires very long-term management because some of the hazardous radioactive material present takes hundreds of thousands of years to decay to a safe state. This raises both practical and moral concerns. In order to keep the waste from causing harm on a geological timescale, surface storage, the current interim solution, cannot be seen as a permanent answer. Storage deep in the ground offers an inherently more stable environment for the waste with a greater fail-safe capacity than surface storage. All nations have taken the view that deep storage is the only viable long-term option. However, scientific study based on storing waste this way is still in its infancy. As a result, scientific programmes to investigate possible sites and designs for deep storage can be expected to take decades to complete and are not assured of success. This means that knowing whether deep storage would be safe, let alone deciding where facilities could best be located, may be a long way off. In the UK, the management of nuclear waste has been repeatedly identified by expert committees and citizens' juries as a problem which must be resolved before the extension of nuclear power can be contemplated. Yet it remains unresolved, and is likely to be so for some considerable time.

Nuclear waste also raises the moral question of whether it is right to leave as a legacy large and unknown costs and risks for the future. Many would argue that it is not, and that by using nuclear generation to reduce the risk of climate change, we are replacing one set of environmental and social risks with another risk that a vast number of future generations will have to take without, of course, having had any input into the decision-making process. In effect, we would be entering into a Faustian pact on their behalf.

Cost

A further important factor in considering the future role of nuclear power is its costs. One thing is certain: the optimism that the industry expressed in claiming in its early years that nuclear electricity would be 'too cheap to meter' has proven seriously misplaced. The true cost to date has been obscured by decades of government support and subsidies, some of which are hidden and hard to identify. Between 1974 and 1998, OECD countries (the industrialized nations) spent around £100 billion in today's money on support for nuclear research. In the UK in the 1990s, the industry received almost £8 billion in subsidies via the fossil-fuel levy. In 2002, the government provided the industry with an emergency loan facility of hundreds of millions of pounds. Even given all these past subsidies, the running costs of nuclear power are such that there is no likelihood of any further power stations being built in the UK on a commercial basis – further huge subsidies would be needed. Optimists in the industry suggest costs could fall with a new generation of nuclear power stations but, as has been noted, optimism about nuclear power has a poor track record.

Despite the massive public investment in nuclear power, the really big costs are still in the future. No one knows with any certainty what the expense of decommissioning nuclear power stations and storing the waste safely for tens or hundreds of thousands of years is likely to be. One estimate suggests it will cost £85 billion to dismantle the UK's existing nuclear power stations. This is almost 10 per cent of the UK's annual GDP, or £1,500 per person. Another estimate suggests the cost could be more than £24 billion for Windscale alone. Experience of cost escalation in the nuclear industry (Sizewell B cost three times its original estimate) makes it seem likely that these figures will turn out to be substantially lower than the actual cost. Even considering just the financial costs of nuclear power (and not the cost of damage to health and environment it causes), the case for expansion at present is poor.

Future prospects

Given its expense and attendant risks, will nuclear power be a large contributor to supplies of 'zero carbon' energy? Public mood in the UK at present would strongly oppose its expansion and the government has not so far encouraged the industry to believe that it has a future. Construction of new nuclear power stations in the next decade or two is unlikely, unless governmental priorities change considerably, the public can be persuaded of the case for nuclear power, and large amounts of subsidy money are provided – three unlikely prospects. There are many cheaper, risk-free ways of reducing carbon dioxide emissions and these are far preferable to the nuclear power option.

Energy efficiency

Energy efficiency is regarded as a 'good thing' by economists, engineers and environmentalists alike. It offers reduced energy use through technological improvement, without having to change business practices, behaviour or lifestyle. It focuses on 'win-win' solutions with outcomes that are better for the business balance sheet or consumer budget and better for the environment. Although efficiency is less glamorous, less visible and less symbolic than renewable energy, it is a major focus of many UK and EU policies and in theory has much to offer. And there is plenty of evidence to show that energy efficiency achieves what it promises. However, improving energy efficiency in isolation does not necessarily lead to energy savings because other factors, such as increased travel or higher ownership of central heating, are working in the opposite direction, to increase energy consumption.

This section describes the achievements of energy efficiency in the domestic, transport and business sectors so far. In addition, the potential for further efficiency improvements as well as the limits and limitations of efficiency are explored. The fundamental question is whether energy efficiency can deliver overall energy savings rather than just reduce the amount of energy needed for a particular activity, beneficial though that is.

Overview of energy efficiency

There is no overall measure of energy efficiency for the economy.* Instead, there are many different measures in different sectors – from the kilometres per litre for new cars, to the centimetres of loft insulation installed in the housing stock. These measures are not linked: one cannot act as an indicator for the others. So, assessing energy efficiency and its prospects in detail requires analysis of many different technologies for all three of the main sectors of use – domestic, transport and business.

There is a range of views on the prospects for future improvements in efficiency. Some commentators are extremely optimistic. For example,

* National energy intensity, which is primary energy divided by GDP, is sometimes used as an indicator of efficiency. However, it is affected by many factors, from climate and population density to the structure of energy supply and the economy. Therefore it is not a true measure of efficiency and is not used here.

books have been written about both 'Factor 4' and 'Factor 10', offering potential four- or ten-fold increases in the efficiency with which energy and other material resources are used. Even the less optimistic would agree that, despite decades of efficiency improvement, there is still considerable scope for improvement of many key energy-using technologies. A recent EU-wide study suggested it would be possible to reduce greenhouse gas emissions from 1995 to 2010 by 16 per cent across all sectors of the economy, largely through implementing efficiency measures. Studies from the 1970s to the present day that have looked at the potential savings from energy efficiency often identify savings of around 20–30 per cent within twenty years: as with oil, there seems little immediate prospect of the energy efficiency potential running out.

Domestic sector

Considerable increases have been made in energy efficiency in individual homes and domestic equipment since the 1970s. However, as explained in Chapter 3, there has been a failure to save energy overall due to higher standards of comfort demanded, increasing levels of electricity use, and growing numbers of households. On its own terms, energy efficiency has been successful, and it can be argued that, without energy efficiency, energy use would have been very much higher than it is today. Indeed the Building Research Establishment suggests that UK domestic energy consumption could have been almost double what it was around the turn of the twentieth century if energy efficiency measures had not been introduced from 1970 onwards. However, from a climate change perspective, the only things that count are the carbon and energy savings that have been achieved *overall*. At a sector level, energy efficiency has not delivered these.

Space and water heating

Most carbon dioxide emissions from existing houses come from the energy used in heating houses. Consumption is determined both by how effectively they retain heat and by the efficiency of the heating system (and, of course, by people – their attitudes, preferences and income). Both factors have improved in parallel to give much improved efficiency overall. The insulation standards of buildings have improved over time: the average new home built in 2002 requires only 60 per cent of the heating energy of the average existing home. This improvement has been achieved through successive rounds of Building Regulations. The efficiency of gas boilers, used by most householders for space and water heating, has also increased from around 65 per cent twenty years ago to an average of 75 per cent today, with the best new boilers being over 90 per cent efficient – a figure that is unlikely to be much improved upon.

Lights and appliances

The efficiency of fridges, freezers and fridge-freezers, washing machines and light bulbs has also improved considerably and some progress has been made with tumble dryers, dishwashers, televisions and VCRs. Most of this efficiency improvement has been in response to EU legislation or voluntary industry agreements, and supported by the action of UK institutions such as the Energy Saving Trust. The appliances for which there has been less progress include digital receiver/decoders, cooking appliances, and the use of stand-by power in many gadgets (discussed in Chapter 3).

Prospects for further efficiency increases

Many experts believe that further improvements in efficiency will lead to more energy saving in the domestic sector. The Energy Saving Trust has calculated that a quarter could be saved by 2020. Research carried out by Oxford University estimates possible carbon savings of 17 per cent from 1998 to 2020 for domestic energy use, excluding space heating, even allowing for growth in the standards of service. Most encouragingly, a study from Leeds Metropolitan University states it would be possible to reduce carbon dioxide emissions by 60 per cent in the UK housing stock by 2050 largely through efficiency measures. All of this research is based on careful energy modelling and implementation of technologies which currently exist at costs that are affordable. Thus the indications are that very vigorous policy action combined with public and private investment in efficiency could lead to considerable carbon and energy savings in the housing stock.

These savings come from a wide variety of efficiency improvements in both new and old houses. In new housing, the energy demand for heating could be dramatically reduced: there are international examples of housing using as little as 10 per cent of the heating energy of new UK houses. Older houses too could be improved. Experience shows that comprehensive improvements in efficiency achieve a 50 per cent reduction in energy use. Energy-efficient lights use only a quarter of the energy of their conventional rivals and the most efficient new fridges use just half the energy of others on the market. With most household equipment, the greatest challenge is not to improve the technology further, but to ensure it is the most efficient version of a product that is purchased. With new housing, minimum standards can be raised through building regulation while new mechanisms must be found for upgrading the existing housing stock and for financing improvements, in some instances subsidizing them. In summary, the key to increased efficiency is a more stringent environmental policy in which efficient houses, light bulbs and boilers are eventually the only ones available to consumers.

However, despite the huge promise of greater energy efficiency in the UK domestic sector, past experience must be borne in mind. The considerable improvements in efficiency since 1970 were not accompanied by matching energy savings. The danger is that new ways of using more energy will outstrip the benefits of efficiency gains. Efficiency can only guarantee conservation as the end result where overall energy saving is an explicit political goal. As the next section will show, experience from the car industry is not encouraging.

Transport sector

Transport is the sector in which energy use is rising most rapidly, despite considerable improvements made in the design of conventional vehicles. (Alternative vehicle technologies such as electric cars are discussed separately in a later section.)

Efficiency of energy use in the transport sector is a more complex subject to understand than in the domestic sector. It can relate to vehicles, passengers or journeys – and each category is measured differently. The three categories are:

- **Vehicle efficiency:** The efficiency of moving the vehicle around, measured as kilometres per litre (or miles per gallon).
- **Passenger / freight efficiency:** The efficiency of moving the passenger or goods around by that mode of transport, measured as energy per passenger kilometre or for freight per tonne kilometre.
- **Journey efficiency:** The fuel required to get from A to B by any method of transport.

Vehicle efficiency is the easiest of the three to measure and can be used to set improvement targets for vehicle manufacturers. Passenger / freight efficiency is usually a more useful measure because it includes more of the complexities of transport systems, such as the occupancy rates of public transport or loading rates of lorries. In comparing the energy use of cars and buses, it is essential to look at energy use per passenger, rather than per vehicle. Finally, journey efficiency captures another important determinant of energy used in transport – the mode by which people or goods travel. A change from cycling to school to taking the bus represents a reduction in journey efficiency because the traveller has changed from personal (renewable) energy to using fossil-fuel energy.

Some progress has been made in improving vehicle efficiency for cars: the average fuel consumption of a new petrol, two-wheel drive car fell from 9.3 to 7.9 litres per 100 kilometres between 1980 and 1987 – a 15 per cent reduction – but did not improve from then to 2001. The benefits from the efficiency increases due to improved engines were offset by greater vehicle weight and the provision of additional, energy-hungry features such as air conditioning. In addition, as car ownership has increased, car occupancy levels have fallen. Thus energy consumption per passenger kilometre has increased and passenger efficiency has decreased. In addition, walking and the use of bicycles, motorcycles and public transport have fallen – all of these are lower energy users than the car, so journey efficiency has decreased even further.

There is little evidence of significant improvement in vehicle efficiency for either freight vehicles, buses and coaches or trains. By contrast, the vehicle efficiency of aircraft has improved, as has the efficiency per passenger kilometre. In a special report on aviation, the IPCC estimated that efficiency per passenger kilometre for *new* aircraft has improved by around 70 per cent between 1950 and 1997.

To summarize, significant transport efficiency gains (for vehicles, passengers and freight) in recent decades have been limited, and those that have occurred have been massively outweighed by the increasing use of transport. All the journey efficiency trends have been in the wrong direction, with the private car being increasingly favoured even for very short journeys and aircraft used for domestic inter-urban travel far more than in the past.

Future prospects

For cars there remains considerable scope for improvement in vehicle efficiency. The European Commission has negotiated a voluntary fuel economy agreement with manufacturers that should reduce average carbon dioxide emissions from new cars to 25 per cent below 1995 levels by 2008. So far, indications are that the target will be met, although much of this is due to higher sales of diesel cars, which are more efficient than petrol ones. Moreover, given the average car occupancy of 1.6 people, increasing the number of people per car and hence reducing emissions per passenger kilometre is a possibility. On the other hand, with current policies, car ownership is expected to keep rising, which will lead to falling occupancy levels and hence increased emissions per passenger.

The vehicle efficiency of buses, trains and lorries is not expected to improve much in the near future. However, management practices can lead to reductions per passenger kilometre in the use of public transport vehicles by raising occupancy levels; freight vehicles can also be made more efficient by minimizing 'empty running' and by using vehicles better matched in size to the loads carried.

There are prospects for further improvements in efficiency in aircraft, both through vehicle efficiency and improvements in occupancy rates. Larger aircraft may also prove to be more fuel-efficient. British Airways has led the industry in setting a target of 30 per cent reduction in fuel consumption per passenger kilometre over a twenty-year period. But even this rate of improvement would only offset at best a quarter of the expected 5 per cent annual growth in flights. The efficiency potential both per aeroplane and per passenger kilometre is very small in comparison with the expected growth in demand.

Looking at the broadest definition of efficiency, that of the energy used in movement from A to B, there is still much scope for improvement. For passenger transport, a quarter of car journeys are less than two miles, a distance easily covered by most people by bicycle or on foot. Some road freight could move back to rail. Air travel could be replaced by rail for many European journeys. There are many other possibilities, but rather than improvements in efficiency, they rely on behavioural change which people have generally been unwilling to make – so far. It would seem that improving efficiency will not achieve lower overall energy consumption in the transport sector. That will require more fundamental changes.

Industrial and service sectors

The industrial and service sectors together use just over one-third of the UK's energy, so improving efficiency could make substantial savings. Energy uses in these sectors vary considerably between different types of businesses and organizations. For many small businesses, energy use is concentrated mainly on space heating and cooling, lighting and IT equipment. At the other end of the scale, considerable amounts of energy are used in specialized industrial sectors such as steel making and chemicals. Generally, businesses where energy use is a large part of the running costs have improved efficiency for their own financial benefit. They can afford to employ specialized energy managers and are alert to potential savings. It is the smaller enterprises, where energy costs are low, which tend to have taken less action, whether through lack of knowledge, time or capital for investment.

In the late 1990s, the advent of 'e-business' led to hopes that economic activity could be 'dematerialized' – that is, that wealth could be created without significant material resources or energy input. The classic example is selling a music information file, which can be stored on the purchaser's existing computer, instead of having to manufacture and sell CDs. However, such neat substitutions of information goods for physical goods are the exception rather than the rule. In most cases, business activity has a material basis, and most successful e-businesses make their money from selling conventional goods, as well as highly energy-intensive services such as flights. E-business and internet shopping have not so far proved to be routes to more energy-efficient business practices.

Future prospects
There are still many opportunities for improving efficiency, from better house-keeping and from installing new technologies and management systems. Potential savings of between 1 and 30 per cent have been identified in different branches of industry, with around 20 per cent being a commonly available reduction. In general, the percentage of savings available tends to be lower in the more energy-intensive industries, where energy makes up a large part of the cost of production, and higher in less energy-intensive sectors, such as in the retail business, where energy costs represent only a very small percent-age of turnover. Many of the savings can be made with little or no investment of capital, as they rely on improved housekeeping and better control of exist-ing systems. Given that these sectors of the economy are such major users of energy, it is essential that they make their contribution to the overall reduction in carbon dioxide emissions. This may involve changing the types of products they make and the services they provide, and not just changing their current practices and procedures.

The future role of energy efficiency

It is important to recognize the limitations inherent in the use of energy efficiency as a strategy for energy saving. First, the opportunities for energy saving afforded by energy efficiency are based on the gradual replacement of hundreds, thousands or millions of pieces of energy-using equipment, typi-cally over at least a ten-year cycle. Three-quarters of UK housing in 2050 already exists – only one-quarter will be built from 2002 onwards. Thus the more efficient new housing will have a limited impact on overall energy use. Similarly, a new design of aircraft might take ten years in development and another ten in construction. Consequently new, more efficient aircraft could take at least twenty years to come on to the market, and it will take far longer to replace the existing stock of aircraft. In nearly all cases, the prospects for greater energy efficiency have to be seen in this light.

Second, savings from efficiency will become progressively difficult to achieve over time, as the best opportunities are taken up first. Allied to this, improvements in efficiency are also progressively difficult to achieve (as with time reductions in athletic events). For some domestic equipment, such as washing machines and gas boilers, the most efficient models now on the market would be hard to improve upon.

Third, there is considerable debate about whether energy efficiency is used as a way of gaining more energy services rather than reducing energy use. In other words, the extent to which owning a more efficient piece of equipment simply allows the owner to use it more than he or she would have before. This concept is known as 'take-back'. The evidence from the domestic sector indicates that where there was unmet demand prior to the introduction of more efficient equipment, there is a higher degree of take-back. One important example is space heating. Not surprisingly, evidence shows that the colder people's homes are prior to efficiency improvements, the more the gain in efficiency taken back as increased warmth (that is, additional energy use). Another example is the car: as cars of the same engine size have become somewhat more efficient over recent years, people have bought larger, higher-performance vehicles and therefore with higher energy consumption. The result has been that the overall efficiency of the average petrol car purchased has not improved over the past fifteen years. The existence of take-back, in variable degrees, does not invalidate claims for energy efficiency in principle, but it does require that projected savings are reduced by the degree to which take-back is expected to occur.

Finally, by making running costs of energy-using equipment cheaper, energy efficiency itself is in some cases contributing to rising expectations and 'needs'. 'Need' is not a fixed standard, it is socially and culturally determined, with yesterday's luxury fast becoming today's essential. Heating the whole house, rather than just the living areas, has become common practice because of the availability of more efficient central heating systems and better-insulated houses. In this, energy efficiency is part of the broader technological and economic advance which is serving to bring energy-using equipment and activities (such as cars, central heating systems, long-distance holidays) within reach of most British people. It is very difficult to disentangle the role of energy efficiency in constructing these energy consumption 'needs', but there is little doubt that it plays a part in this process.

Efficiency is probably the most effective approach to reducing carbon dioxide emissions but it cannot guarantee savings under conditions of continuous growth of energy-based activity and the very powerful forces promoting this. It has been seen that energy efficiency has not been sufficient to counteract these. In general, far greater success has been achieved in finding new ways of using more energy than in implementing energy efficiency measures. Some would argue that all that is required are more vigorous, comprehensive and better-funded energy efficiency programmes. While agreeing that far more could be done to improve energy efficiency, the evidence to date makes it difficult to believe that efficiency on its own can deliver significant, sector-wide savings in energy.

Alternative vehicle technologies

There are a number of alternative vehicle technologies that may offer ways of reducing carbon dioxide emissions. These include electric vehicles, 'hybrid' electric vehicles, and vehicles powered by LPG, natural gas or fuel cells. The interest in these alternatives is in part a response to the fact that the opportunities for increasing the efficiency of conventional vehicles, while retaining their current size, power and features, are limited. However, research is also motivated by other issues, such as concern about local air quality (and pollutants in car exhaust fumes, such as sulphur and nitrous oxides).

Electric vehicles

Electric vehicles have been under development for more than a hundred years, yet the same major disadvantage, that of not being able to travel very far on a single charge, has persisted despite improvements in battery technology. The other difficulty they face is in delivering high enough power for acceleration. Given current performance expectations, these are serious barriers to a greater uptake of electric vehicles. Apart from milk floats (around 18,000) and other specialist vehicles, there are just a few hundred electric vehicles operating on the roads of the UK – mainly cars and delivery vans operating in urban environments. The best prospect for electric vehicles remains as urban conveyances used for short journeys made at low speed. There is little prospect of electric vehicles replacing conventional ones.

The carbon dioxide emissions of electric vehicles are determined by the type of electricity used to charge their batteries. For this reason, an electric car using conventionally produced electricity could increase emissions compared with an equivalent petroleum-fuelled car. Renewable electricity is needed for 'zero carbon' operation, but as shown earlier, supplies of this electricity are currently very limited, meaning that electric vehicles are unlikely to constitute a mass-market solution for the foreseeable future.

Hybrid electric vehicles

Much more promising are hybrid electric vehicles. These have both a fossil-fuel internal combustion engine and an electric motor powered by batteries. There are different degrees of hybridization, ranging from simply capturing energy lost during braking and returning it to the battery (regenerative braking), to full hybrids which allow periods of electric-only operation and have an extended battery range. The electric fuel system is used at speeds lower than 10 mph and for stop-start driving, whereas the fossil-fuel engine is used outside urban areas and to travel at higher speeds. The energy efficiency of the car increases with the degree of hybridization, with up to 28 kilometres per litre (80 miles per gallon) being possible in a mid-size car with full hybrid features, representing a saving of around 50 per cent of energy and carbon dioxide emissions on current conventional cars of an equivalent size.

Several hundred thousand hybrid cars are in operation around the world, mostly in the USA and Japan, where they were developed partly in response to government requirements for less-polluting vehicles. Two hybrid cars, the Toyota Prius and Honda Civic, have recently come on sale in the UK. Their fuel consumption (and carbon dioxide emissions) are about one-third lower than current equivalent petrol cars. As these cars do not yet employ the full complement of hybrid technologies, further progress on their energy efficiency is expected. The range available should increase as many major car manufacturers have announced their intention of introducing hybrids within the next few years. Although the prospects for their improvement and uptake seem promising, vehicles using this advanced technology are currently relatively expensive.

Fuel cell powered vehicles

Car manufacturers are investing considerably in research and development on fuel cell vehicles and there has been much hype about the prospects for combining the fuel cell with the use of hydrogen as an environmentally safe transport solution. This is because, in the long term, renewably produced hydrogen could allow road vehicles to operate with zero emissions.

Fuel cells could indeed provide energy cleanly and efficiently. Like batteries, they produce electricity by converting energy by a chemical reaction directly into usable electric power. But unlike a battery, a fuel cell has an external fuel source, typically hydrogen gas. Inside most fuel cells, hydrogen from a fuel tank and oxygen from the air combine to produce electricity and warm water. Fuel cells are efficient, portable converters of fuel to electricity, capable of turning 50–70 per cent of the energy in hydrogen fuel into electricity.

The advantage of fuel cells depends on how the fuel they use is supplied in the first place. There are three main options, in descending order of the reduced levels of carbon dioxide emissions they offer:

- Pure hydrogen produced by using renewable energy.
- Pure hydrogen produced by using fossil-fuel energy.
- Fossil-fuel, which is used to create hydrogen on board the vehicle.

Only using hydrogen produced by renewable energy offers zero emissions. As will be seen in the next section on the 'hydrogen economy', prospects for producing large amounts of hydrogen from renewable energy lie at least thirty years into the future. The two other options offer smaller yet important advantages over conventional vehicles. Using fossil fuels to power a fuel cell could offer the same scale of emissions reductions as a hybrid electric vehicle – that is, up to 50 per cent. So although fuel cell vehicles have potential advantages, the technology is still in development, and they are therefore not expected to be commercially available for at least ten years.

Alternative hydrocarbon vehicles

The popularity of both natural gas and LPG as alternatives to petrol and diesel in a conventional engine has been growing in the UK. The key reasons are the low rate of duty on these fuels and the availability of government-funded grants for adapting vehicles to LPG. This makes them significantly cheaper to run than their petrol or diesel equivalents. However, the government announced in summer 2003 that it was reviewing preferential support for these fuels and vehicles, the outcome of which is likely to affect their future adoption.

LPG is used mainly in cars and light vans, whereas natural gas tends to be used in lorries and buses. There are around 25,000 LPG vehicles in the UK at the moment, mostly owned by companies. Refuelling has become easier because there are now over 1,100 public sites supplying this gas in the UK, often located at petrol stations. Almost all LPG vehicles sold in the UK are bi-fuel, allowing the driver to change from LPG to petrol and vice versa. Natural gas vehicles are operated by many local authorities and companies. Because natural gas has to be compressed or liquefied for use in engines, it requires a heavy storage tank to keep it in this state. That is why the majority of natural gas vehicles are heavy-duty trucks and buses, as larger and heavier fuel tanks pose less of a problem for these vehicles. Natural gas is unlikely to be used in cars.

A recent study by the Institute for Public Policy Research (IPPR) showed that emissions from LPG vehicles were about 15 per cent lower than from the petrol equivalent, and emissions from vehicles using compressed natural gas 25 per cent lower. This comparison is based on greenhouse gas emissions per kilometre, on a 'well to wheel' basis, which includes emissions from all stages of fuel extraction, processing, delivery and usage. However, ordinary diesel-fuelled vehicles generate emissions around 20 per cent lower than the petrol equivalent. The government's Vehicle Certification Agency concurs with this analysis, suggesting that LPG vehicles tend to fall between petrol and diesel vehicles on carbon dioxide emissions performance and that compressed natural gas is on a par with diesel. Thus LPG and natural gas do offer some advantage over petrol vehicles, but neither provides better carbon savings than a switch to diesel. Although these fuels may affect local air quality less, they do not represent a significant lowering of carbon dioxide emissions compared with vehicles currently in use.

Future prospects

Alternative vehicles still retain the size, status, power and features of conventional cars. Even the best of the available high-tech options offers little more in savings than those from choosing a smaller car or a diesel rather than a petrol engine. In the Ford range, for instance, a small car (Fiesta) emits just over half the carbon dioxide per kilometre compared with a large one (Mondeo). This is almost identical to the Vauxhall range, where its small car (Corsa) has an energy efficiency in excess of 18 kilometres per litre in comparison with the larger one (Omega) which achieves less than 9 kilometres per litre. In addition, compared to petrol, diesel vehicles have significantly lower carbon dioxide emissions per kilometre (up to 20 per cent lower) because of the higher efficiency of the diesel engine. Hence more efficient vehicles with lower emissions and requiring no new technology are already available.

Electric cars and fuel cell cars can guarantee zero emissions only if there is a plentiful supply of renewable electricity, which is not likely to be available for several decades. LPG and natural gas offer relatively small savings. The most promising near-term technology is the hybrid electric vehicle, which already has proven carbon-saving advantages over a conventional car. If it reaches its potential, it could save half of the energy per conventional car (assuming there is no increase in weight and energy-using gadgets, such as air conditioning). Even combining the best available technologies with smaller vehicles will not be enough to avoid having to take further action in the light of the predicted increasing ownership and use of cars. In other words, technology alone cannot provide the level of reduction in carbon dioxide emissions that we need.

The hydrogen economy

In a speech on sustainable development made in 2003, the Prime Minister said: 'Hydrogen holds out the potential to replace fossil fuels, especially in transport, and could transform our energy system – offering a vision of a transport system that is completely clean and with no exhaust emissions.' Hydrogen could also replace natural gas for heating and hot water in homes, offices and industrial buildings. However, this vision is perhaps more a product of wishful thinking than of a realistic analysis of the current situation. At the moment, hydrogen is used as a fuel in just a small number of demonstration projects. Further major technological advances and infrastructure changes would be required for an economy based on energy from hydrogen to be established in the UK. So how much faith should be put in the hydrogen dream?

Producing hydrogen

Hydrogen is in many respects the ideal fuel. The energy released per tonne in combustion is more than twice that of any hydrocarbon and no carbon dioxide is produced because no carbon is involved. However, hydrogen does not exist abundantly in nature so, like electricity, it must be produced in the first place – using energy – which is when carbon dioxide emissions may arise. Hydrogen has two very important advantages over electricity. First of all, hydrogen can be stored fairly simply, holding the energy until it is needed. Second, it can be used as a portable fuel in an internal combustion engine. Neither of these is true, by and large, for electricity.

As with electricity, hydrogen is only as 'carbon free' as the energy used to produce it. There are three main ways of doing this:

- Hydrogen can be chemically separated from hydrocarbons, such as methane (natural gas), using a device called a reformer.
- It can be created by the splitting of water into hydrogen and oxygen using electricity (electrolysis).
- It can be manufactured biologically via photosynthesis or fermentation.

The world's chemical industries already produce millions of tonnes per year of hydrogen from hydrocarbons, but this results in carbon dioxide emissions. On the other hand, *if* the electricity for electrolysis is derived from renewable sources, the emissions are truly zero, as water is the only combustion by-product. However, as explained earlier, renewable energy provides only a small percentage of UK electricity and the future prospects of generating it to a significant extent from nuclear power are poor. Thus this is a big 'if'. An alternative future is one in which the Sahara Desert is covered with solar panels, generating electricity which in turn produces hydrogen to be liquefied and then shipped to the UK. The prospects of this visionary solution being realized are remote.

Using hydrogen

The most promising use of hydrogen is in transport applications, particularly in cars and other road vehicles. There is little immediate prospect of using hydrogen in aircraft because liquid hydrogen requires around four times the storage volume of kerosene (the fuel used in planes) for the same energy content. Also, the water vapour emitted at high altitudes could add to the greenhouse effect. Indeed, the RCEP concluded that hydrogen 'does not offer an attractive alternative for fuelling aircraft'.

If it is assumed that hydrogen could be produced using zero-carbon electricity for use in road vehicles, then two key challenges still remain: distribution and on-vehicle storage. While its energy content on a mass-for-mass basis is better than petrol, hydrogen is more difficult to transport and store because it is a gas, not a liquid. A hydrogen gas fuel tank that contained a store of energy equivalent to a petrol tank at atmospheric temperature and pressure would be more than 3,000 times larger than the petrol version. So hydrogen would have to be stored in a compressed or liquefied form on board the vehicle and during transport to filling stations. There are several technical hurdles to be overcome before this could be easily and economically achieved. Alternatively, hydrogen could be created on board the vehicle from natural gas using a 'reformer'. But carbon dioxide would still be emitted and so this option offers no advantage in terms of reducing emissions when compared with conventional, diesel-fuelled vehicles (as outlined above).

Future prospects

In the long term, hydrogen produced from renewable sources could allow road vehicles to operate with zero 'well to wheel' carbon dioxide emissions. Hydrogen might also replace other fuels, such as natural gas, in the home. However, the situation at the moment is that renewables are only a very minor component of electricity production in the UK, making the prospects for widely available renewably generated hydrogen somewhat distant. A recent study concluded that it would not be beneficial in terms of reduction of carbon dioxide emissions to use electricity based on renewables to produce hydrogen for use in vehicles, or elsewhere, until there is a surplus of renewable electricity. Higher carbon savings will be achieved through displacing electricity from fossil-fuel power stations. That surplus is unlikely to be available for at least thirty years, if then. Thus, while research into the hydrogen economy and associated technologies may lead to worthwhile advantages in the longer term – say, 30–50 years ahead – in the short term it does not offer a solution to climate change.

Carbon sequestration

Carbon sequestration is the permanent removal of carbon dioxide gas from the atmosphere so that it no longer contributes to the greenhouse effect. It is different from the other technical options examined because it offers the possibility of continuing to use fossil fuels with no net carbon dioxide emissions to the atmosphere (that is, carbon sequestration could fully compensate for the emissions produced by burning fossil fuels).

There are two types of sequestration. Trees and other plants and organisms absorb carbon dioxide from the atmosphere to grow and they incorporate the carbon within their own molecular structure – this is biological sequestration. Industrial processes can also be used to capture carbon dioxide, which can then be compressed and stored beneath the ground – this is geological sequestration. Biological sequestration via afforestation – growing trees to 'mop up' carbon dioxide from the atmosphere – is a well-known concept that has been included in the Kyoto Protocol. It is already being used both in the UK and internationally to 'offset' carbon dioxide emissions from fossil-fuel use. Geological sequestration, on the other hand, is a more recent idea. It includes storage underground and under the sea. However, sequestration in both its forms remains controversial, principally owing to growing scientific doubts about its long-term effectiveness. The question that must be answered is whether it can make a significant contribution to a low-carbon future.

There is also another type of biological sequestration – in oceans. This is being investigated by some researchers, but the risks and benefits are far from agreed. One idea for this form of sequestration is to add iron sulphate to sea water to increase the growth of algae, which take up carbon dioxide from the atmosphere when growing and will deposit the carbon at the bottom of the ocean when they die and sink to the seabed. However, many marine scientists regard this solution as highly risky for other ocean life and too unproven to be worth further consideration at present.

Biological sequestration: tree planting and forest management

At present, there is a great deal of political and commercial interest in the idea of planting or restoring forests to offset carbon dioxide emissions. Indeed, tree planting, which has been powerfully symbolic of environmental improvement for many years, would appear to be an ideal solution to climate change. However, carbon storage within forests is a very complex subject which is not yet fully understood. This makes carbon savings from biological sequestration a less reliable proposition than simply using less fossil fuel.

Forests have a key role in the global carbon balance, and they can act in three different ways. These are as follows:

- **Carbon reservoirs:** Global forests contain around four-fifths of the carbon stored in land vegetation. Of this total, about 60 per cent is held in tropical forests with the rest divided between temperate and boreal forests, mainly in the higher latitudes of the northern hemisphere.
- **Carbon sinks:** Forests, soils and other vegetation currently absorb about 40 per cent of manmade emissions. There are two reasons for this. First, forests and soils are recovering naturally from past damage, and vegetation is regrowing and absorbing carbon dioxide. Second, there has been a global acceleration of photosynthesis (which governs the rate at which plants absorb carbon dioxide from the atmosphere) due to increasing levels of carbon dioxide in the atmosphere and, in some areas, nitrogen deposition. But as temperatures rise further, carbon uptake will be reduced, offsetting this trend.
- **Sources of greenhouse gases:** Deforestation mainly in tropical regions and changes in land use cause approximately one-fifth of global warming.

Planting trees to offset carbon dioxide emissions requires long-term commitment. A tree plantation will absorb carbon dioxide as the trees grow, but eventually the growth rate and absorption of carbon dioxide slows until in a fully mature plantation the rate of tree growth and carbon sequestration are close to zero. Old trees die and release their carbon back into the atmosphere as carbon dioxide, while young trees grow and absorb carbon in roughly equal measure. If the sequestration benefit is to be maintained (for ever), there are three options. First, the mature plantation can be kept alive indefinitely and protected from fire and pest attack. Second, it can be harvested but the harvested material must be stored away from the atmosphere in perpetuity, for example by incorporating it in building construction. Third, the plantation can be harvested and the plant material burnt as an alternative to fossil fuels. These are stringent conditions.

The UK potential

The total area of woodland in the UK has been expanding and now covers nearly 12 per cent of the landmass. However, the total carbon stored in the country's vegetation, 80 per cent of which is held in these woodlands, is equivalent to around only two-thirds of the UK's carbon dioxide emissions for *one year*. Moreover, given current plans, additional tree planting will make only a small contribution to offsetting these emissions: annual carbon sequestration from all the additional trees planted in the UK between

1990 and 2020 is expected to offset less than 1 per cent of today's emissions.

The Tyndall Centre (Britain's leading centre for climate change research) estimates that the carbon stored from doubling the UK forested area over the next fifty years would be equivalent to only 2–3.4 per cent of the current annual UK emissions of carbon dioxide. Their 'conservative achievable' scenario, which involves a lower rate of forest expansion, gives an annual offset of 0.7–1.3 per cent. Put another way, to offset total UK emissions at the current rate for the next fifty to a hundred years would require tree planting over an area *four times* the size of the UK. This is clearly impossible. The UK land area available for forestry is such that, even with ambitious forestry expansion plans, national forest planting cannot conceivably soak up our current levels of carbon dioxide emissions to any significant degree. Carbon sequestration to offset a significant proportion of UK emissions would therefore have to be located abroad. So what are the prospects for tree planting on a grand scale overseas?

The global potential
The land surface available for forestry growth and restoration in other countries of the world is considerable. A recent review by the Royal Society, the UK's foremost scientific institution, concluded that the potential for increasing biological carbon sequestration by 2050 could result in making one-quarter of the reductions in carbon dioxide emissions required by then to avoid serious damage from global warming. However, it suggested that there is 'little potential for increasing the land carbon sink thereafter'. So although this global potential is important, even if the maximum amount of savings could be achieved, it would be insufficient to avoid taking substantial additional action to reduce the impacts of fossil-fuel usage.

There are moral implications too. Paying other countries, nearly all in the developing world, to plant and maintain forests on behalf of the UK (or any other country) would mean that the donor country would be able to 'cherry pick' cheap forestry projects to reduce its own emissions. The country that had created the new forest would be left with more expensive options for achieving its own carbon reductions. Another concern is that the social and environmental consequences resulting from large-scale afforestation projects, including lowering water tables, planting exotic monocultures and restricting traditional livelihoods, might be overlooked, leaving these countries with further problems. Fundamentally, it seems right – and has been the position of the EU in the Kyoto negotiations – that developed countries should make the majority of their carbon reductions through actions within their own borders. So although biological sequestration could play an important role globally, wealthy countries such as the UK should not be entitled to rely heavily on paying other countries to plant or restore forests on their behalf.

As the Royal Society concluded in its report on the subject: 'There is still considerable uncertainty in the scientific understanding of the causes, magnitude and permanence of the land carbon sink.' In addition, it seems that the contribution that changes in forestry can make to carbon sequestration is limited in relation to the size of the problem. The first priority must be to try and prevent further destruction of current forests, a complex issue which has defied resolution despite considerable international attention over the years. For the UK, any serious contribution from forestry sequestration would have to come from forestry overseas, which is ethically questionable. And clearly, enhancing carbon sequestration through afforestation does nothing to reduce carbon dioxide emissions at source, or to restructure the economy for a low-carbon future.

Geological sequestration

Geological sequestration consists of first capturing carbon dioxide, then compressing and transporting it, and finally storing it permanently underground. Technologies for capture, compression and transport of carbon dioxide are already available. However, research on the safety and permanence of underground storage is in its early stages and has raised some questions which have not yet been answered but are critical to its wide application.

It only makes sense to capture the carbon dioxide at points where large quantities are produced – for example, power stations. The concept of abstracting it from a power plant's exhaust gases has been considered since the 1970s, but it is only in recent years that government and industry have pursued the idea seriously. The process available can capture around 90 per cent of the gas but energy is used in the process, giving an overall reduction in emissions of about 80 per cent compared with the same plant without carbon capture. The recovered emissions must also be compressed prior to transport and this too requires further energy input.

The final stage involves storage in large natural reservoirs that currently contain saltwater (saline aquifers), gas, oil or coal (hydrocarbon fields) or salt (salt caverns). The majority of sites are below the sea: given the asphyxiant properties of carbon dioxide, it is hard to imagine storage taking place below land. Assuming that well-sealed sites can be identified, the key risk is that unexpected pathways for migration will lead to the carbon dioxide making its way back to the surface. In hydrocarbon fields, old drilling holes, as well as natural faults, would provide an additional risk of slow but steady escape.

There is only one industrial-scale example in the world of this technology being used. Since 1996, about 1 million tonnes of carbon dioxide have been injected into an aquifer formation almost 1 kilometre below the bed of the North Sea at the Norwegian Sleipner gasfield. This represents about 3 per cent of Norway's carbon dioxide emissions for one year. The facility is being very carefully monitored. Storage either within hydrocarbon fields or salt caverns has also been proposed but has not yet been tried out.

Carbon sequestration could be a very important technology for fossil-fuel companies as a way of maintaining their energy markets while at the same time meeting targets for carbon reduction. However, the risks of undertaking underground storage are only just beginning to emerge. For nuclear waste, which becomes less hazardous with time as the radioactivity decays, the designers of a potential underground test laboratory had to quantify the risks of radioactive waste escaping over the next 100,000 years. Carbon dioxide will not become less dangerous with time. This raises the question of what the guaranteed storage time should be, if geological sequestration were proposed to deal with a significant proportion of emissions. Based on present knowledge, geological sequestration seems an expensive and risky option. Even if safe storage underground could be made reliable over the very long term – a very difficult condition to meet – it seems highly unlikely that it can be a major part of the solution to climate change.

Conclusions

The three carbon-reducing options that technology offers are as follows:

- Making more efficient use of energy.
- Using less carbon-intensive energy.
- Capturing and storing the carbon dioxide emissions from using fossil-fuel energy.

The question at the start of the chapter was whether, in combination, these could deliver a 60–80 per cent reduction in carbon dioxide emissions by 2050. The answer is that they could not: efficiency has practical limits and a poor track record of delivering *actual* savings; acceptable energy sources for electricity not based on fossil fuels are limited, and are almost non-existent for heating and transport fuels; and carbon sequestration within the UK can make only a small and possibly unreliable contribution. The hydrogen economy dream has little basis in reality.

However, our conclusion is not that technological change cannot help reduce the impact of energy use, rather that it is unlikely to do so sufficiently in a 'business as usual' world where forces for growth will always dominate.

What is required is a reducing limit on the amount of carbon dioxide we in the UK are permitted to emit (a theme that is explored in Chapters 7 and 8). Then all the technological options would play a strong role in helping the country to live within its 'carbon budget' while protecting the environment. Energy efficiency and renewable energy will only be able to play their full role in supplying energy services with lower impacts under a regime of reducing energy demand. Otherwise their notional energy and carbon savings will continue to be largely irrelevant against a background of ever-increasing energy use and carbon dioxide emissions.

Part Three:
How We Can Save the Planet

7. The Solution
Fair Shares: The Only Way

We have seen in the preceding chapters both the scale of the crisis we are facing and some of the strategies that have been proposed for dealing with it. This chapter sets down in some detail the only solution that will work. It is practicable, equitable, credible and assured of success. Before describing it, however, it is useful to recap briefly the conclusions of the earlier chapters in order to appreciate why the solution must be based on a framework approach.

Chapter 2 explained the evidence for climate change and the serious concerns raised by a 'business as usual' development path. The effects of climate change are already with us. If we carry on as we are, average temperatures could be 6°C higher by the end of the century. Chapter 3 showed how energy use, the main cause of global warming, is expected to increase both in the UK and worldwide. It pervades all aspects of modern life and there are many forces operating to lock people into patterns of activity that are not otherwise possible; changing these will require collective action. Chapter 4 elaborated on the mental defences put up in order to avoid confronting the awesome reality of climate change and our responsibility for it. Chapter 5 demonstrated that current government action is only scratching the surface of the problem. The UK is on far too slow a path to a sustainable, low-carbon future. Chapter 6 has shown that technological solutions, although having considerable potential, will not be sufficient to reduce carbon dioxide emissions to the extent needed to prevent dangerous climate change. The conclusion is that we need to think beyond energy efficiency and renewable energy, and towards ideas of sufficiency, of social and institutional reform and of personal changes that incorporate *much* less energy and lead to *much* lower emissions.

Global, national and personal solutions are vital. Global agreement is vital because the target of a 60 per cent reduction in emissions by 2050 proposed by the UK government, and the 80 per cent minimum that the authors of this book consider more realistic, only works to limit climate change sufficiently if all countries of the world are also engaged in emissions control and have their own reduction targets. The UK cannot act unilaterally, although this does not mean that it cannot take the lead. But, equally vitally, people within the UK must be engaged in the project – the government cannot do it without us and technology will not provide the magic fix. This means devising a national scheme to share out the UK's allocation of carbon dioxide emissions. Both global and national approaches are suggested in this chapter, based on political realism and principles equity and effectiveness.

The global solution

A global solution requires global agreement. It is widely acknowledged that the Kyoto Protocol, the first international agreement on greenhouse gas reduction, will deliver only modest savings in global emissions. It is intended to be the first of a succession of treaties. Future treaties will need to involve all countries of the world, not just the developed countries currently committed to reductions under Kyoto. This means agreeing a framework for a global sharing of the finite capacity of the atmosphere to absorb greenhouse gases without serious damage to the climate.

This section presents a framework for such an agreement. It is the one thought by an increasingly influential number of national and international institutions to be the most promising approach for global negotiations. Before describing the approach, its fundamental basis is outlined.

Morality and the meaning of fairness

Climate change is an ethical issue and tackling the carbon dioxide emissions that cause it is a moral imperative, in all likelihood the key one of our time. The full effects of today's emissions will show up in the future as increased temperatures, storms, drought, flooding, loss of biodiversity and stress on human populations – in some instances extreme. The immediate effects are already highly disturbing but not so spectacular. However, preventive action needs to be taken now. Not to do so would burden future generations with the consequences of our irresponsible actions.

Equity issues are central to the international climate change debate. This is for both principled and practical reasons. Intergenerational equity is at the heart of policy on reducing greenhouse gas emissions because, as the emissions accumulate in the atmosphere for hundreds of years, much near-irreversible harm has already been caused and our current emissions are accelerating the process. The principle underlying this approach is the same as the ideal of sustainable development which was expressed in the Brundtland Report of 1987, *Our common future*, as 'development which meets the needs of the present without compromising the ability of future generations to meet their own needs'. Equity is key for practical reasons as well. Without equity, transparent in its application, there can be no realistic prospect of public acceptance or political agreement to introduce the measures needed. Not only should issues of fairness be the basis of the response to climate change, they are an essential component of the process.

However, there is more than one definition of equity. Deciding what equity means is crucial to answering very real and practical questions such as who is allowed to suffer how much climate change damage, and who gets to emit how much carbon dioxide. Equity can be interpreted in three ways:

- People have equal rights to use the atmosphere, and so should receive equal allocations of carbon dioxide emissions.
- The greater the capacity to act or the ability to pay, the greater should be the contribution to the solution (somewhat like the 'logic' used to share emission reductions between the EU countries).
- Polluter pays – the greater the contribution to the problem, the greater should be the contribution to the solution.

Each of these definitions of equity would result in different allocations of responsibility for achieving carbon reductions and, indeed, the second two would require a regular re-allocation of responsibility over time. Although over the short term both definitions have their attractions, a principle that helps determine how global emissions should be shared in the long term is needed and the equal rights argument seems by far the simplest and most morally defensible option. Therefore the definition of equity used in this book is that of equal rights to use the atmosphere. We believe too that this is the only one which can be successful in international negotiations. As it happens, the richest countries that have the greatest capacity to act are the ones which not only have been responsible for historically high levels of emissions but are also currently the most highly polluting. Thus they are the ones who should and will have to make the greatest changes under an equal rights framework.

Description of the global solution: contraction and convergence

A brilliant, imaginative and simple means of reaching a just global agreement on emission reductions has been put forward. Known as contraction and convergence (C&C), it was first proposed by the Global Commons Institute (GCI) in 1990. Recognition of its unique qualities as a framework for combating climate change has grown at an astonishing rate since that date. C&C is founded on two fundamental principles: first, that the global emission of carbon dioxide must be progressively reduced; and second, that global governance must be based on justice and fairness. However, this latter requirement has not been included for moral reasons alone; the GCI also claims that it would be essential for getting agreement from developing countries to take part in global emissions reduction. Its phrase 'equity is survival' encapsulated the point that there can be no global security unless climate change is restricted to a manageable level, and this cannot be achieved without all countries of the world sharing this common objective.

C&C consists of:

- **Contraction:** An international agreement is reached on how much further the level of carbon dioxide can be allowed to rise before the changes in the climate it produces become totally unacceptable. Once this limit has been agreed, it is possible to work out how quickly current global emissions must be cut back to reach this target.
- **Convergence:** Global convergence to equal per capita shares of this contraction is phased towards the contraction target by an agreed year.

C&C is a set of principles for reaching agreement. In fact, it simplifies climate negotiations in a remarkable way to just two questions. First, what is the maximum level of carbon dioxide that can be permitted in the atmosphere? Second, by what date should global per capita shares converge to that level? Using C&C does not entail a particular concentration of carbon dioxide emissions as being the safe limit, nor a timescale for reductions. Based on current projections, the GCI suggests that it would be irresponsible to adopt any concentration higher than 450 ppm. However, it acknowledges that the 450 ppm target might well have to be revised downwards in the light of new evidence.

The GCI argues that C&C offers a realistic 'framework' to replace the 'guesswork' involved in the Kyoto Protocol. The targets in the Kyoto agreement are not based on any reliable understanding of the safe, or at least not-too-dangerous, limits of greenhouse gases in the atmosphere. Rather, the reductions agreed were determined by what was considered to be politically possible at the time of the negotiations between the thirty-seven countries involved. By contrast, C&C would use the best current scientific knowledge to set maximum levels of carbon dioxide emissions in the atmosphere, and hence maximum cumulative emissions. While the date of convergence would be subject to agreement, the principle of equal rights for all would remove the potentially endless negotiations that would otherwise occur, with each country making a case that its contribution to global reductions should be modified in the light of its special circumstances.

Another important element of the C&C proposal is the ability of countries to trade carbon emissions rights. Countries unable to manage within their agreed shares would, subject to verification and appropriate rules, be able to buy the unused parts of the allocations of other countries or regions. The lifetime of permits would be restricted (to, say, five years) to discourage futures speculation and hoarding. Sales of unused allocations would generate purchasing power in vendor countries to fund their development in sustainable, zero-emission ways. Developed countries, with high carbon dioxide emissions, would gain a mechanism to mitigate the expensive premature retirement of their carbon capital stock. They would also benefit from the export markets for renewable technologies that this restructuring would create. At the same time, the application of the C&C proposal would not only have the virtue of making a major contribution to shrinking the gap between rich and poor, both within and between countries, but also encourage the adoption of types of energy with low carbon dioxide emissions.

What would C&C look like?

The impact of C&C on the emissions allowances for people from different countries can be envisaged by reference to the scenario illustrated below (Figure 8) in which the limit on carbon dioxide in the atmosphere is set at 450 ppm and convergence is achieved by 2030.

The C&C graph shows how levels of carbon dioxide emissions related to fossil fuels have evolved over time for six blocks of countries: the USA; other OECD countries (which includes all the EU and other European countries, Australia, New Zealand, Japan and Canada); the remaining countries of the former Soviet Union (FSU); India; China; and the rest of the world. Not surprisingly, most of the historic carbon dioxide emissions, prior to 2000, are the responsibility of the developed world. After C&C is introduced, for instance in 2000 in this scenario, there is a period of adjustment up to 2030 by which date equal emissions rights have been achieved. The graph assumes that there is no trading between countries; in reality, the pattern of emissions

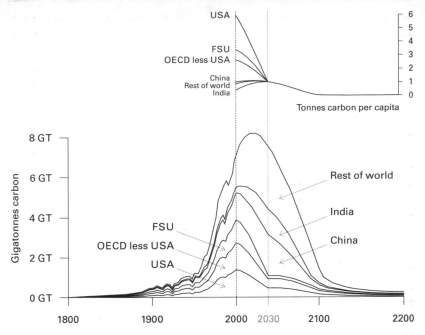

Figure 8: Carbon dioxide emissions under C&C (shown gross and per capita) for a maximum of 450 ppm atmospheric concentration achieved by 2100, with 'permits' for per capita emissions converging to equality achieved by 2030 (source: GCI website)

might be rather different from this, with rich countries emitting more, having paid the poorer countries for the privilege of doing so.

The graph shows how per capita emissions of carbon dioxide would change under this C&C scenario. The highest carbon-emitting countries have to make the largest contributions to the overall reduction in emissions, so the change per capita required is greatest for the USA, followed by the FSU countries and then the OECD countries (including the UK). Emissions from developing countries would be permitted to increase to 2030. Thereafter, as for the developed countries, their emissions allowances would gradually reduce over time to ensure that the 450 ppm target was not breached.

International support for C&C

A large number of national and international bodies have endorsed C&C as the right way to tackle climate change and have published statements supporting it as the framework for negotiations. Since 1997, these have included key government spokesmen in China, India and the Africa Group of

Nations to the UN climate negotiations. It has been endorsed by most European environment ministers and by the overwhelming majority of MPs in the European Parliament. Other organizations lending their support are as diverse as the UK Chartered Insurance Institute, the worldwide membership of the UNEP Financial Initiative, and two of the world's largest insurance companies.[*]

Numerous NGOs and academic institutions have also backed it, resolutions of the UK branch of the United Nations Association have been passed backing it, and the Intergovernmental Panel on Climate Change Third Policy Assessment has acknowledged its logic. The World Council of Churches and other religious organizations have called for a commitment to the framework, and the World Bank has published statements recommending C&C as the basis for effectively and equitably reducing greenhouse gas emissions.[†]

National support for C&C in the UK

The attitude to C&C in the UK is especially interesting. Numerous highly placed individuals and institutions have specifically endorsed it. Many of these have also taken steps in public to advocate C&C to the government. Prominent among them are the Greater London Authority, the Performance and Innovation Unit (PIU) in the Cabinet Office, the IPPR and the RCEP.

In 2000, the RCEP report *Energy: the changing climate* vigorously restated the case for contraction and convergence set out by the GCI at the UN climate negotiations in 1996. Its chief recommendation to the government was that the UK should adopt the principle of C&C. It also urged the UK government to lead the diplomatic efforts to establish it as the basis of future international agreement. This was echoed a year later in the PIU energy report to the government.

The RCEP report gave several examples of C&C in scenarios demonstrating different rates of sharing international control of future greenhouse gas emissions. This was to show first the relationship between different rates of the contraction of greenhouse gas emissions and the consequently different levels of atmospheric concentrations, and second, the negotiability of the rate of convergence to equal per capita shares of the global contraction process.

Greatest emphasis was given to the so-called 'CO$_2$ Doubling Scenario' in which the upper limit of carbon dioxide concentration in the global atmosphere does not rise above 550 ppm. In other words, in this scenario emissions are controlled globally so that concentrations rise to no more than twice the

---

[*] See details on the GCI website: www.rcep.org.uk/pdf/chp4.pdf
[†] See details on the GCI website: www.gci.org.uk/consolidation02.html

level of carbon dioxide present in the atmosphere prior to the Industrial Revolution.

The scenario requires a *global* cut or 'contraction' of emissions of over 60 per cent against 1990 levels within a timeframe of just over one hundred years. In the scenario, the RCEP also recommended that the international convergence to equal per capita shares of this global contraction budget should be complete by the year 2050 and that international shares consequent on these rates (or any rates) of C&C should be internationally tradable.

In doing so, the RCEP formally joined in advocating the need for international equity whereby those *developed* countries with the high per capita emissions most responsible for causing climate change should immediately begin to experience a reduction of 'permits' to emit while, at the same time, the *developing* countries with low per capita emissions would simultaneously experience a continued increase in 'permits' to emit. In creating and making permits tradable in this way, the RCEP accepted that the C&C mechanism could achieve some redress for developing countries which have been least responsible for the emissions causing climate change while also being most vulnerable to its consequences.

Additionally, in line with the C&C model, the RCEP recommended that after 2050 the international shares of the contraction budget should remain proportional to populations at that time – with 2050 as a base year – as global emissions continued to fall in line with the requirement not to exceed the doubling of concentrations of carbon dioxide in the atmosphere by 2100. In the detail of this scenario, the RCEP also pointed out that this particular combination of rates for C&C would entail a 60 per cent cut in the UK's emissions permits by 2050, falling to a cut of more than 80 per cent by 2100. The RCEP further urged the government to adopt this path for UK emissions control.

UK government response to C&C

The government framed its response to the RCEP in its Energy White Paper published in February 2003. In this, the Prime Minister called for an international 'Climate Covenant' involving all nations – including the USA – acting together to avoid the dangers of 'a planet ravaged by climate change'. He also announced that the government accepted the RCEP recommendation to reduce carbon dioxide emissions in the UK by 60 per cent by 2050.

When it was pointed out to the government that this figure was derived from the RCEP's C&C scenario, the government published a memorandum stating that their approach was actually based on C&C. However, they held short of overtly advocating C&C to the international community, implying,

wrongly in our view, that there are viable alternatives to C&C and that more
time was needed to find them.

What alternatives have been proposed?

A review of proposals for tackling climate change published by a leading UK
think-tank, the New Economics Foundation, concluded that the GCI's C&C
strategy is the only one which offers assurance of, first, arriving at a defined
atmospheric concentration; second, the equitable allocations that developing
countries have rightly stated to be an essential part of any agreement; and,
third, the potential for immediate implementation.

The two other main approaches are the 'Brazilian' and the 'Kyoto plus'
proposals. They actually create more problems than they solve, however. The
Brazilian Proposal (formulated and proposed by the Brazilian Ministry of
Science and Technology in 1997 during the negotiations on the Kyoto Protocol) emphasizes responsibility for the historic emissions that have caused the
rise of atmospheric concentrations, temperature and environmental damage.
In this, countries with a longer history of industrial development would bear
a greater share of responsibility than those with shorter histories. Thus, with
this greater share, the UK would face a 63 per cent reduction by 2010 against
1990 levels, while Japan's reduction would be less than 10 per cent. This puts
all the responsibility for emissions reduction on the older developed countries,
excludes developing countries from quantified commitments, and has no
formal concentration target. Thus the proposal contains an explicit escape
clause for some developed countries, notably the USA, which argues that it
will not be the signatory of any international agreement on climate change
which does not involve a commitment from these other countries.

By contrast, as its name applies, the 'Kyoto plus' proposals are variations
on the theme of continuing the existing approach. However, they lack a target
for reducing concentrations of carbon dioxide in the atmosphere and a clear
idea of where the process ought to be going. Thus, neither of these alternatives
provides the same vital steps of, first, fixed atmospheric concentration targets;
second, an agreed timescale for implementation; and third, and at least as
important, a clear constitutional framework for developing countries' partici-
pation in the C&C process.

Why C&C is the right approach globally

C&C is the only approach that addresses questions of equity and emissions
reduction targets that are scientifically determined. It ensures that an acceptable level of carbon dioxide emissions per capita is agreed upon and main-

tained throughout the world. The approach allows the developing countries to increase to the equitable level emissions resulting from their economic growth, while developed countries are compelled to adopt environmentally sound pathways to economic growth and policies that will reduce emissions to the level required. Climate change threatens potentially catastrophic losses. C&C mitigates this by integrating the key features of global diplomacy and development necessary for long-term prosperity and security.

When tested as viable alternatives to C&C, the other approaches are feasible only when seen as facets of C&C. Thus the clarifying simplicity of C&C makes it possible to secure a global deal that is at once comprehensive and comprehensible.

How could it come about?

A framework based on C&C requires international agreement and political consensus. Although the Kyoto Protocol has turned out to be disappointing, there are good precedents for effective global action on environmental problems. One key example of this is the 1987 Montreal Protocol on reducing substances depleting the ozone layer which has been ratified by most countries. The other is the Convention on Long-range Transboundary Air Pollution – designed to abate airborne emissions causing acid rain, eutrophication and ground-level ozone – which is supported by forty-nine parties, including the EU. There is no doubt that climate change is a more serious and difficult problem than either of these. Nevertheless, experience shows that, with the will to act, collaboration at the international level on critical environmental issues can succeed.

Recognition of the need for action on climate change should grow steadily as evidence of the damaging consequences of inaction accumulates. Some commentators suggest that disasters will be necessary before countries act. Sadly, with the WHO estimate of 160,000 people dying from global warming each year in developing countries, as cited in Chapter 2, and almost 15,000 people dying from excessive heat in France alone during the summer of 2003, disasters are already happening and close to home. Even though the heatwave in France cannot be proven absolutely to be the result of climate change (due to the variable nature of weather), it is the type of event that is predicted to occur with increasing frequency as the climate changes. For many European countries, the very dry and hot weather of summer 2003 provided an unsettling glimpse into the future. The weather led to extensive forest fires, huge damage to agriculture, losses in the tourist industry and numerous deaths. Some nations are already finding the increasing evidence of the effects of climate change harder to ignore.

In the UK, the government should take the additional step of declaring that it wishes future climate treaties to be based on the C&C principle. It has already gone part of the way by choosing its 60 per cent reduction target by 2050. Now it needs to give public backing to C&C and use its leadership in climate negotiations to encourage others to do the same. This position requires all-party consensus as climate negotiations are going to feature on international political agendas for decades, beyond the life of particular governments.

What is the UK solution?
Personal carbon rations

Assuming the UK takes part in a global agreement to reduce carbon dioxide emissions, how can a reducing emissions quota be shared out? Based on the equity principle embodied in C&C, the answer suggested here is for a system of personal 'carbon' rationing for the 50 per cent of energy used directly by individuals. Indeed, as part of a global agreement, per capita rationing would be the obvious mechanism for all countries. Although the word 'rationing' may be offputting, associated as it is with images of deprivation and restriction, it is used deliberately here to be very clear about the idea being presented and to tap into the collective consciousness of national action for the greater good. The words 'allowance', 'entitlement' or 'domestic tradable quota' are alternative descriptions of the same concept.* The main features would be as follows:

- Equal rations for all individuals, with minor exceptions.
- Year-on-year reduction of the annual ration, signalled well in advance.
- Personal transport and household energy use to be included.
- Tradable rations.
- A mandatory, not voluntary arrangement.

In the following pages, the concept is described more fully, followed by a more detailed justification. Chapter 8 discusses further the implications of such a policy for individuals, business and society as a whole.

Equal rations

Clearly, giving people equal carbon rations – an equal 'right to pollute' – is equitable in theory and reflects the international equity principle embodied in

---

* The concept was originally proposed by Aubrey Meyer (founder and director of the GCI) and Mayer Hillman. It was also developed by David Fleming (of the Lean Economy Institute) and is now the subject of detailed research by Kevin Anderson and Richard Starkey (of the Tyndall Centre for Climate Change Research).

the C&C proposal. There may have to be some exceptions to this rule. However, in general, it will be better for society to invest in provision for their energy efficiency and use of renewables so that they can live more easily within their ration rather than to keep tinkering with the ration. The largest group likely to face problems are those living alone, because they have higher energy consumption per capita than those living in multi-person households. At the same time, the more exceptions which are granted, the lower will have to be the ration for the rest of the population. It must be best to keep the system as simple and transparent as possible, and pay more attention to the help people can receive in adapting to a system based on an equal ration rather than increasing it for them.

Reducing rations

Carbon rations will have to decrease over time, in response to the need to reduce global emissions and to allow for a rise in population. By giving due warning of the reduction in the future ration, people would be able to adapt their homes, transport arrangements and general lifestyles at the least cost and in the least disruptive way to them individually. Experience in the industrial sector shows that industry has been able to produce more efficient equipment (such as fridges and washing machines) at no extra cost if allowed time to adapt the design and manufacturing processes. The same is likely to be true of people adapting to lower-energy, more carbon-efficient lifestyles.

Personal transport and household energy use

By including personal transport and household energy use, half of the energy-related emissions of carbon dioxide in our economy would be covered. The other half comes from the business transport, industry, commerce and public sectors which produce the goods and services we all use. In theory, it might be possible to manage this half by calculating the 'embodied' emissions in each product or activity (such as the emissions used to produce the apple, stereo equipment or car) and give consumers a further allowance to be used when buying products. However, this would be both extremely complex and data-intensive as well as being very difficult to apply to some goods and services – how could you 'carbon rate' a haircut or a hospital stay? It would be much simpler to make the non-domestic sector directly responsible for reducing their share of carbon dioxide emissions. Thus a separate scheme is envisaged for non-personal energy use (see below).

Tradable rations

Not everyone will need to use their full carbon ration. Those who invest in efficiency and renewables, and who lead lives with a lower energy input, will not need all of their ration and will therefore have a surplus to sell. Those who travel a lot, or who live in very large or inefficient homes will need to buy this surplus to permit them to continue with something like their accustomed lifestyle. Thus people will want to trade carbon rations. Price would be determined by the availability of the surplus set against the demand for it. For this purpose, a 'white' market would be created, possibly via a government clearing 'bank' or maybe a version of the electronic auction system 'eBay' – called 'Cbay'?

In the early years of its introduction, relatively wealthy people will be able to buy the surplus from those who lead energy-thrifty lifestyles. But the cost of doing so will rise steadily as the system of rationing is progressively introduced. Economic theory says that by allowing trading, any costs of adapting to a low-carbon economy will be kept to a minimum. In addition, clearly there would be informal bartering – for example, people accepting lifts with others are likely to offer to pay some of the carbon 'cost' as well as contributing to the cost of the petrol used.

A mandatory approach

In order to be effective, carbon rationing would have to be mandatory. A voluntary approach would not succeed: individuals would be unwilling to start taking action for the common good unless they saw others doing likewise, and the 'free-rider' would have far too much to gain. In circumstances such as this, when the wider public interest is at risk and the issue one of critical importance to the welfare of the community, government intervention is necessary. Equally, a voluntary alternative to carbon rationing would be unlikely to make significant savings, as recent history suggests that appeals to reason and conscience have not been effective in achieving major changes in our irresponsible patterns of consumption.

Administration

Administration of carbon rationing should be simple. Each person would receive an electronic card containing that year's carbon credits (see illustration below). The card would have to be presented on purchase of energy or travel services, and the correct amount of carbon would be deducted. The technologies and systems already in place for direct debit systems and credit cards could be used.

There are relatively few sellers of gas, electricity, petrol, diesel and other fuels, and flows of fossil fuels are already very well recorded and tightly regulated in our economy. Introduction of such an allowance scheme therefore would affect few businesses, and those involved would be large businesses with the capability to adapt. In addition, given the existence of the 'white' market that we propose, there should be little opportunity for a 'black' market to develop. To ensure that the introduction of carbon rations would not be too complex, we suggest that initially it should exclude journeys by public transport as these account for only a small part of road transport emissions.

Beyond the individual

One objection to personal carbon rationing is that it is unfair to put all the responsibility for reducing carbon dioxide emissions on individuals when they have limited power to act. Their emissions are determined by many factors and some – such as the 'carbon intensity' of electricity, the choice of lower-carbon fuels available, the types of efficient cars and equipment on the market – are not necessarily within the power of individual to change. They should not be alone in attempting to reduce their personal carbon dioxide emissions. Rather, society and the economy must enable, support and widen their opportunities for doing so.

There are a number of social, technical and policy innovations which would be needed to make it possible for people to live within their carbon allowances. On the technical side, these could include 'smart meters' which informed people how much of their carbon ration for that year was left, which appliances were using most energy, how much carbon could be saved, for example, by reducing the time spent in the shower, or by only heating bedrooms in the late evening. Alternatively, energy companies could install in houses sophisticated carbon management systems which took these decisions automatically and guaranteed carbon savings for customers. In terms of policy, equipment that used less energy could be favoured through devices such as VAT, labelling, minimum standards and subsidy.

At present, the purchase of the most efficient types of equipment is encouraged, whether cars, refrigerators or washing machines. In future, the emphasis will be on items using the lowest amount of energy or with the lowest emissions, with much better information available at the point of purchase of everything which uses energy, from new and existing homes to televisions and mobile phones. It would become in the economic interest of manufacturers to supply goods which make the lowest use of carbon. Socially, one would envisage that values would gradually change so that thrift rather than profligacy in energy and carbon become valued.

Learning from experience

There has been no recent experience of long-term rationing of energy (other than by price) in the UK. The nearest comparison is food rationing introduced at the beginning of the Second World War when the civilian consumption of food, clothing and other goods was reduced drastically as these commodities were in short supply and economic resources were prioritized in favour of the war effort. The food element of the ration covered meat, bacon, cheese, fats, sugar and preserves in fixed quantities per head. The principle of a flat-rate ration for all, which ignored the diverse needs of manual workers at one extreme and small children at the other, was justified since not all foodstuffs were rationed. In addition, it was recognized that certain categories of the population had special nutritional requirements and therefore other schemes were superimposed on this common basis. Nevertheless, equity was a key feature of its introduction and maintenance up to the early 1950s.

Although food rationing may evoke images of hardship, and therefore may be thought to be a poor exemplar for personal carbon allowances, it is important to remember that rationing, coupled with subsidies and price controls, promoted social equality, and consumption became more equal in contrast with the intense inequalities that had existed previously. Despite difficulties, contemporary opinion polls showed that there was general support for rationing and food control. The British experience with food rationing is that the chosen scheme was seen as fair and retained public support. (In fact, overall nutrition and health also improved.)

In comparison with food rationing, carbon rationing would in some respects be less prescriptive and intrusive in everyday life. People could select from a range of ways in which to adjust their lifestyles and energy use in order to reduce their personal carbon dioxide emissions. However, the need for carbon limitation is likely to be less clearly felt than the need for food rationing. This was necessary to ensure that the population remained well fed at a time of national crisis and restricted food supplies. If society had not accepted rationing and, associated with that, price controls, many people would have gone hungry or even starved – an obviously unacceptable outcome.

In the case of climate change, the increasing effect of carbon dioxide emissions on people's lives would not immediately be felt. The motivation for the UK population undertaking carbon rationing as a whole would be different, and the personal connection with the benefits of carbon rationing would be less obvious. We can therefore learn from looking at food rationing, and be encouraged that it was successful, though clearly there are many differences between that experience and what will be necessary in implementing a system of carbon rationing. As we have already argued, education has a key role to play so that the public understands why rationing is being introduced and for that reason supports it as the only fair and realistic way of responding to climate change.

How could this come about?
Political consensus

A cross-party consensus on C&C and the adoption of carbon rationing is necessary. The future of the planet is too important an issue to be treated as a political football and placed in practice, though not in the rhetoric, fairly low down on the electoral set of priorities, as it is now. This must change. It would be devastating if there were no common purpose and instead political groupings vied with each other to obtain electoral support by making less demanding commitments on climate change in their manifestos. However, the likelihood of achieving such co-operation is by no means remote – it is just that it has not yet been sought. None of the main UK parties has expressed reservations about either the significance of climate change as one of the most hazardous environmental issues ever to face human society, or the need for serious, concerted action to limit its impacts. Their positions are described briefly below.

- The present Labour government's concern is convincingly reflected in the lead it took in the Kyoto negotiations in 1997 and the Prime Minister's statement at the G8 Summit in 1998 that 'climate change remains the greatest threat to our prosperity'. More significantly, it is reflected in the government's adoption of a target of a 20 per cent reduction in UK carbon dioxide emissions by 2010 and a 60 per cent reduction by 2050.
- The Conservative Party also supports the ratification of the Kyoto Protocol and the responsible action that the UK government is taking in international negotiations on combating climate change.
- The Liberal Democrats acknowledge that a stable world climate is under threat and that more effective action than has been taken to date is called for – according to the principles of contraction and convergence – with the long-term goal of equalizing per capita emissions across the world.
- The Green Party has set the issue of climate change at the heart of its manifesto. It also supports the C&C framework, and has called for an 80 per cent reduction in carbon dioxide emissions by 2040 – a significantly more ambitious target than that of the Labour government.

Education and persuasion

Education will be vital to help to break the cycle of denial concerning the gravity of climate change and to persuade people of the necessity for action. Broad education policy at all levels of society should be the primary instrument for this process, involving all political parties, industry, financial institutions, civil society, and bodies with moral authority, including, of course, the churches. There has never been a greater need for organizations with very different agendas to work together. That may seem a long way off. But it can change rapidly as more politicians and influential individuals declare their position on the issue and acknowledge the unsustainability of current policies and practices.

Although the number of individuals who are active in demanding a proper response to climate change is relatively small at present, a domino effect could easily occur. Some environmental campaigns have been very effective, usually without any political affiliation, in striking a chord with a growing proportion of the population over the last thirty years. Most tellingly in the context of climate change, their appeal has been greatest among the younger generation – those who will have to pay the higher price for this generation's dilatoriness.

Educators will need to work more closely with schools on environmental issues relating to climate change, which must feature more prominently in the National Curriculum. These could form the main part of citizenship classes where children would learn and understand the importance of sustainability, its relationship with their lifestyles and that of their parents, and its role in ensuring the future integrity of the planet.

The media will also have a role to play. However, at the current time, a great deal of advertising revenue for television, newspapers and other outlets is derived from cars, air travel, foreign holidays and other 'high carbon' products and services. Given this conflict between the public interest and their own commercial self-interest, the media are unlikely to lead the way to a low-carbon future.

Introducing a UK scheme

Introducing carbon rationing in the UK does not rely on a prior global agreement on a C&C basis. It could precede the necessary international global carbon reduction treaty. Carbon rationing could be used in parallel with existing government policies to move towards the current domestic targets of a 20 per cent reduction of carbon dioxide emissions by 2010 and 60 per cent reduction by 2050. Once the system was established, rations could be reduced if necessary as a result of the global agreement which demanded further and faster cuts in emissions. As well as leading the world in introducing long-term carbon reduction targets, the UK could demonstrate an approach that would deliver these reductions equitably and effectively.

If not this, then what?

The main alternative to personal carbon rations is carbon taxation. There is support in some policy-making circles for this option. Like carbon rationing, taxation could provide a cap on energy consumption, but in this case by pricing. Six European countries have already introduced carbon taxes (although some of these taxes are little more than traditional energy taxes under a new guise) but there has been a long history of failed attempts to introduce an EU-wide carbon tax. As most EU countries are not on course to meet their Kyoto targets, current carbon taxes have not reduced emissions sufficiently. For this reason, the level of taxation required to do so is unclear.

In any case, there are a number of problems with taxation as the route to curbing emissions. First, it does not have the same moral basis as rationing or the same psychological resonance as the principles contained in the global contraction and convergence approach, for it allows those with higher incomes to pollute more as of right. Second, its effectiveness varies according to the trade cycle – a tax rate that achieves its objective in a period of strong economic growth will be much too harsh when that same economy is in recession. By contrast, rationing has the advantage of certainty of result: it is clear exactly what carbon savings will be made. Finally, because of the nature of taxation, it would not engender the same 'all in it together' social cohesion towards the goal of lower emissions as would rationing based on equal carbon shares. Without considerable social support for this shared goal, it would be impossible to make the necessary savings.

What is more, carbon taxation would be difficult to introduce for personal energy use. In the UK there would be strong resistance to any form of additional taxation on household energy use. This would stem more from the usual resistance to additional taxes – it has a specific political history. It results from consciousness of the problems of the millions of people in 'fuel poverty', those people who already have difficulty affording adequate energy services, particularly winter heating. Domestic energy taxation is inherently regressive because lower-income households spend a greater proportion of their income on energy than the better-off. As already noted, there was also strong resistance in the UK to further taxes on petrol and diesel in the late 1990s. Neither household nor personal transport energy use would be politically easy to tax further. This is often glossed over by supporters of taxation.

High levels of taxation would be necessary in order to reduce carbon dioxide emissions to the necessary level. Experience of the 'fuel tax escalator' showed that increasing tax on petrol at 6 per cent above the rate of inflation was not sufficient to lower demand significantly. Research in 1999 established that an annual increase in petrol costs of £60, equivalent to only about one week's spending on transport, would have the effect of cutting car use by no more than 0.5 per cent. The RCEP estimated that an emissions charge on aircraft would need to add £40–100 on each fare for it to cover the damage caused. This would of course lower demand but there is every reason to believe that the massive carbon taxation necessary to reduce it by 60 per cent or more would be far less politically or socially acceptable than the incremental introduction of a system of personal carbon rationing.

Winners and losers

A system of personal carbon rationing, based on the principle of contraction and convergence, scaled down year on year, will bring in its wake rising environmental benefits, particularly in terms of limiting damage from climate change. However, its introduction and phasing-in is bound to lead to difficulties for many individuals. Nevertheless, if a world with personal carbon rationing seems unacceptable, just imagine how much less acceptable would be a world without assuredly effective action taken to tackle climate change. This should be the key perspective: the problem of climate change is not going to disappear.

Carbon rationing is not a perfect solution. It will have its losers as well as its winners. Changing patterns of consumption to a low-carbon society will have knock-on effects on the business community as well as on individuals. Industry and commerce will also have to make major changes in its practices under carbon restrictions. Although this book has not specified a mechanism by which carbon allowances and reductions will be introduced for them, their scale of reduction will have to match that for the domestic and personal transport sector. Initially, this is likely to be focused on taking up opportunities for energy efficiency. However, over time, the structure of economic activity will also change, with new opportunities arising as old, energy-intensive activities are phased out.

Required reductions in carbon dioxide emissions will simply drive further changes over and above those associated with current forces such as globalization, use of information technology and the shift to a service economy. Table 3 lists some of the possible 'winners' and 'losers' in an increasingly low-carbon society. It is clear that, under carbon rationing, there will be no shortage of new business and entrepreneurial opportunities.

Table 3: Examples of business winners and losers under a low-carbon economy

Winners	Losers
Manufacturers of efficient appliances, lights, cars, etc.	Manufacturers of inefficient appliances lights, cars, etc.
Construction industry, other than the transport sector	Construction industry for the transport sector
Renewable energy manufacturers, e.g. for wind turbines and solar water heating	Manufacturers and suppliers of fossil-fuel energy stations
Biofuel companies	Fossil-fuel companies
Bus and bicycle manufacturers	Car manufacturers
Organic and other UK farmers	Energy-intensive agriculture
Domestic tourism	Overseas tourism
Bicycle repair shops	Garages and petrol stations
Local shops and businesses	Regional shopping centres
Service and knowledge economy	Short-life goods economy
International communication systems	Airlines
Businesses offering low energy/carbon solutions, e.g. zero-energy homes	Businesses selling high-energy systems, e.g. domestic air conditioning
New technologies such as micro CHP, heat pumps, hybrid cars, airships	Old technologies such as direct electric heating

Many in the 'losers' column, such as some manufacturing and construction companies, are also active in the 'winners' column. For example, those currently involved in new transport infrastructure projects are also involved in building schools and hospitals and in providing housing for the large predicted increase in the number of households in the UK over the next twenty years. BP and Shell, as major suppliers of fossil-fuel energy, also have considerable investment in solar power, and many appliance manufacturers produce both efficient and inefficient products. Some of today's companies are already in a good position to adapt to a lower-carbon environment. Indeed, the history of the development of business and commerce shows that diversification has always been a key to sustained corporate success.

The key concern of industry will be the international effects of carbon reductions. Will the UK be put at a competitive disadvantage if it signs up to a low-carbon future? In the long run, assuming that a C&C mechanism is chosen for global agreement on greenhouse gases, the answer is 'no', as all economies will be working under similar constraints. However, in the convergence period, there may well be international distortions. Nevertheless, if the UK moves sooner than other countries towards a low-carbon economy, this will be of benefit because its businesses will be better able to sell their low-carbon expertise abroad. There can be no future in a high-carbon, high-energy economy (or planet).

Rationing is a serious idea worthy of equally serious consideration. It is not perfect but if there were a perfect solution to the problem of climate change that disadvantaged nobody and required no uncomfortable changes, it would have been discovered before now.

Questions and answers

Over the years, the concept of carbon rationing has been presented by the authors to a wide range of both professional and general audiences. We have set out below, under the headings of 'Political', 'Economic and commercial', 'Social' and 'Practical', the questions that have been raised most frequently and the answers we have given. In our view, the challenge implicit in the questions reflects, albeit tacitly, two unacceptable stances. One is the belief that there must be a better means of achieving the necessary reduction in carbon dioxide emissions (almost invariably unstated and untested). But if carbon rationing is not the right answer, then another equally effective solution is required. We have searched for one that stands up to scrutiny but have failed to find it. Alternatively, the questions reflect the dangerous view that current policy, based on advice and exhortation, and on 'human ingenuity' in the field of technology, will enable us to 'muddle through' as we have in the past, and that it can deliver that reduction without the need to prescribe limits on personal choice.

Political

Q. Why should politicians consider the adoption of carbon rationing when some people dispute the claim that human activity is contributing to climate change?

A. *The weight of evidence on climate change is too overwhelming (as demonstrated in Chapter 2) to justify inaction or to propose no more than a 'no regrets' policy of only taking preventive measures known to be cost-effective. We should stress to politicians that they should not presume the public to be opposed to rationing once the case for it has been made.*

Q. Does not a framework based on contraction and convergence require international agreement, including the involvement of the USA? And is not the whole strategy invalidated if these cannot be assured?

A. *The issue of climate change, and the need to curb the impacts of human activity on it, is certain to rise to the top of the political agenda as evidence of its damaging consequences accumulate. With concern about the progress of climate change growing apace, we can foresee all-party coalitions achieving consensus on the direction that needs to be taken. As far as US involvement is concerned, it may be only a matter of time before its refusal to take part in these negotiations leads to it becoming a* pariah *state on the international field. Moreover, it could be pointed out that the USA will not be spared from the adverse impacts of climate change – these are obviously likely to spur it into action.*

Q. What if the majority of public opinion at a general election or in a referendum is opposed to the introduction of carbon rationing? Are not the prospects of attempting to impose it politically unrealistic – and undemocratic?

A. *Most people do not yet see clearly the links between climate change and their lifestyles. This is largely because they have not been properly informed of the links. Once the connection is made, it will be far more difficult for ignorance to be used as a defence. It will then be more likely that there will be support for a government that was seen to be tackling the problem head on. The argument against carbon rationing based on its lack of attraction to the public at large, and therefore its introduction not being realistic, is by no means as powerful as the one that could be put forward against the dangers of taking insufficient action to prevent the predicted 6°C increase in temperature by the end of the century.*

Q. Surely, no political party will dare produce a manifesto that includes an undertaking to deliver major cutbacks in energy-based activities?

A. *This is an argument for short-term decisions overriding ones that are in the longer-term interest. We should be urging parties to include such an undertaking in view of the consequences for the state of the planet of postponing the inevitable.*

Q. Is such a drastic innovation as carbon rationing – a mechanism more appropriate to a wartime situation – really necessary when people are already doing what they can, for instance, by buying more energy-efficient cars and central heating systems?

A. *The problem is that there is not only an insufficient number of people doing so but also that the opportunities are themselves limited. It is for this reason that we should be calling for the only assuredly effective framework for collective action to be imposed.*

Q. The UK is already doing more towards facing up to its responsibilities on climate change than most other countries. Why therefore should it take the lead in adopting a system of rationing which would then make it uncompetitive in the international market-place?

A. *We are calling for* international *agreement on the adoption of the C&C framework. But, in any case, it would be as inappropriate to take no action on this in the absence of a worldwide consensus as it would be in any situation where an obvious threat to survival was apparent.*

Economic and commercial

Q. Won't the industries with a vested interest in maintaining the status quo, such as the energy-intensive ones of motor manufacturing and international tourism, do all within their power to object strongly to the concept of C&C as its adoption will lead to a steady reduction in demand for their products and services and a rise in their prices owing to the loss of economies of scale?

A. *Yes, this is indeed likely but cannot be considered a reason for inaction. Moreover, the proposal is for a* phased *reduction over a sufficiently long period to ease the transition towards sustainable patterns of activity. To some extent, too, we can be reassured that spheres of the economy that are not energy-intensive – such as education, local shopping, social and leisure activity, non-motorized travel and domestic tourism, and those involved in enabling and promoting energy-thrifty lifestyles – are very likely to prosper.*

Q. Will not shareholders in energy-intensive industries oppose carbon rationing as their primary interest is in increasing profits?

A. In some instances, they may do so. But we would hope that most of them would agree that the sound business principles upon which the profits are made should be compatible with their ethical principles.

Q. How could the adoption of carbon rationing be justified, as it would limit economic growth upon which affluent countries depend in order to be able to afford environmental protection and upon which the survival of Third World economies depends?

A. According to the UK government's Energy White Paper in 2003, a 60 per cent reduction in carbon emissions can be achieved by 2050 at very low cost – equivalent to just a small fraction of the nation's wealth. It is also worth noting that this dangerous counter-argument implies that economic factors override the ecological issue of climate change and its catastrophic impacts. The more important question to ask in this regard is what cost would be incurred if we continue to take insufficient action to limit climate change.

Q. What about the impacts on the economy of carbon rationing by virtue of a steadily diminishing revenue from the reduction in energy-based activities?

A. The Chancellor of the Exchequer has considerable room for manoeuvre in raising taxes on commodities and services on which it is in the public interest to discourage spending – and vice versa. He can therefore raise them on other goods and services on which people will choose to spend their money. Moreover, less revenue will have to be raised as public expenditure in areas such as transport infrastructure will be able to be substantially lowered.

Q. What about job losses in heavily energy-dependent industries?

A. Evidence both of the changes in patterns of employment in the past and of the scope for job generation in industries that will benefit from the adoption of carbon rationing suggests that, while having some social costs, the effects can be accommodated without too much social and economic harm. We can note that, of course, the sooner action is taken, the lower will be these costs.

Q. Many energy-intensive industries are committed to sustainable development and recognize the importance of working towards a hydrogen economy. In these circumstances, is there a need for such a draconian approach as carbon rationing, the imposition of which could devastate the economy?

A. These industries are likely to calculate that their future prospects will be damaged by rationing. In these circumstances, it is in their interest to inspire confidence that, together with government, they are taking steps which will prove sufficient to avoid the need for its introduction. However, the rate of change that is being achieved is far too slow at present (as preceding chapters in this book have shown).

Q. Surely we should be encouraged by the achievements of the energy supply and 'ecotourist' industries in their own operations, the energy-conserving measures they have adopted and their financial support for humanitarian projects and social programmes for local communities?

A. Carbon dioxide emissions from these industries' own operations are typically quoted in their annual 'environmental' statements as is a professed dedication to 'sustainability'. While their achievements deserve to be acknowledged, the extent of the emissions from their customers' use of their products or services is not referred to. The public should insist that these are recorded in their annual statements so that the effects on the environment of the totality and contradictory nature of their activities are transparent. The fact that support is given to worthy causes is irrelevant to the issue of climate change and could be seen as part of a public relations exercise aimed at portraying their apparent concern for the environment.

Q. Rather than adopting carbon rationing, would it not be best to leave the reduction in carbon dioxide emissions, for instance from new cars, to the manufacturers who are already achieving impressive results in improving the efficiency of their products and of their operations on a voluntary basis?

A. Although these improvements are welcome, we should not allow the savings achieved in the use of energy per vehicle to divert attention from the increase in emissions from the rise in the number of cars predicted for future years.

Q. Can't we draw comfort from the growing investment in renewable energy, such as wind turbine and solar projects?

A. Commendable though these projects are, they typically represent only a fraction of the overall expenditure on energy. It is worth noting that even just the increase in the energy industries' investment in exploration for gas and oil far exceeds their total expenditure on their energy-renewable programmes.

Q. Similarly, should we not be encouraged that action is being taken by airport authorities and retailers to limit climate change by investing in new public transport services to encourage people not to travel to airports by car, and basing the design and construction of their buildings upon environmental sustainability?

A. Highlighting these apparent achievements again tends to divert attention from the increasing ecological damage caused by air travel and car use. They should be challenged about the overall impact of their activities. An illustration of this can be seen in the attention drawn to the 'most green' Olympics, held in Sydney in 2000, where visitors were required to rely on public transport and renewable energy was used to meet the demand for electricity. The fact that hundreds of thousands of spectators and competitors travelled by air to and from the event was not mentioned. Another example of this is a major retailer claiming that it has established 'the world's first eco-superstore' which generates its own electricity from renewable sources on site and where the building itself is highly energy-efficient. The paradox is that it serves a substantial population catchment area and most customers drive to the store.

Q. Will carbon rationing not threaten the future of sports and cultural events, such as the Olympics and international conferences, as they entail a significant amount of energy-intensive travel by road, rail and especially air?
A. Yes, it may be that, over time, rationing will lead to the necessity of organizing these events at a regional or local level. These global events will not be able to attract a sufficient number of people prepared to devote so much of their ration to attend them.

Q. With a system of carbon rationing, would not the air travel industry be entitled to preferential treatment with regard to the use of fossil fuels (as there is no alternative to kerosene as a fuel for aircraft), and as individuals cannot reach distant destinations except by air?
A. No. This implies that people who do much flying should continue with this energy-intensive activity. The very reason why we need to discourage air travel is that it is so damaging to the future of the planet. The need to prevent this cannot be compromised.

Q. What about the half of carbon dioxide emissions from the non-domestic sector of the economy that is not contributed directly by individuals? Will it also have to be subject to a system of rationing in which the allocation reduces over time?
A. Yes, an administrative system will have to be set up so that organizations can obtain or are allocated their carbon allowance. They could acquire the units they need from a tender – that is, a form of auction modelled on the issue of government debt. The practical workings of such an allocation mechanism can be left to economists.

Q. Won't this scheme provide more support for the generation of electricity from nuclear power because it is largely carbon-free?

A. It could be favoured for this reason. However, we should be aware of the fact that whether or not nuclear electricity is an acceptable option (as discussed in the previous chapter) is separate from the need to reduce carbon dioxide emissions and should be dealt with accordingly. Personal carbon allowances are designed to help reduce the risk of serious damage from global climate change and not to solve all environmental problems.

Q. Won't new measures of progress based on climate change criteria need to be developed and applied?

A. For sure. Many current measures, such as GDP and the proportion of households that are car-owning, will be invalid. In their place, new ones will have to be developed to indicate the effectiveness of policies aimed at reducing energy use – mileage by what mode, use of bicycles, fuel bills, the life of consumer durable equipment, and so on.

Social

Q. What about the unfairness of a system in which richer people are able to continue with their energy-intensive activities by buying the surplus ration of poorer people?

A. It cannot be denied that this will occur in the early years of the introduction of carbon rationing, but it is not much different from the current broadly accepted incomes gap. Moreover, in contradistinction to this, a system of rationing is socially progressive as it favours those whose lifestyles are low in energy use, generally the poorer members of society. Not only will they be able to increase their incomes by selling their surplus on the open market but, owing to increasing scarcity, its value will rise from year to year as the ration reduces to its base level. It is worth noting that the alternative system of tackling climate change through pricing would obviously have the reverse effect as poorer people would have more difficulty in meeting their energy needs. The question could be turned on its head: we could consider what justification there is for continuing as at present where richer people are able to engage in more energy-intensive activities with relatively little regard to the consequences of climate change.

Q. What about the extenuating circumstances of families whose members are dispersed around the world and wish to see each other when they can afford to do so?

A. Some people have been able to develop very 'global' lives, facilitated by cheap air travel, and to maintain social and family relationships over long distances that were not possible in the past. However, in reality there is little choice if serious climate change is to be prevented. In later decades of this century, when personal carbon allowances are much lower than the initial level at which they would be set, it is unlikely that individuals will be able to undertake much long-distance travel. Adapting to this may be difficult for the small minority of the world's population which currently can benefit from international travel.

Q. What is to prevent couples acquiring additional personal rations by having larger families?
A. *It is proposed that the ration will apply to adults only. Some extra allowance would clearly have to be made for children to cover their additional energy requirements, but this will need to be set so that it does not encourage couples to have larger families. (This is discussed in Chapter 8.)*

Q. Won't the limits on personal carbon dioxide emissions lead to the migration of people for ecological reasons away from parts of the world in which energy needs are exceptionally high and from regions where survival will be critically affected by worsened climatic conditions?
A. *Yes, both instances point to further reasons for us to take the issue of climate change far more seriously than at present. The numbers of the world's population who are likely to be affected look certain to rise sharply and population movements are likely to intensify.*

Practical

Q. Won't a system of rationing and the annual scaling-down of the ration be too complicated for many people to understand?
A. *No, it is very straightforward, comparable with the introduction of wartime rationing in the UK in the autumn of 1939. Much of the adjustment to the declining ration will come about gradually and with considerable advance warning. It will be accommodated through the required learning process from year-to-year experience in the same way that many people in retirement adjust to a declining income.*

Q. What is there to prevent a black market developing and ration books being stolen or easily counterfeited, as occurred during the Second World War?
A. *The proposed system is based on the existence of a 'white' market in which the surplus of individuals' rations can be traded, increasingly profitably as*

the ration reduces annually. That arrangement will remove interest in the development of a black market. As far as theft and forging of ration books are concerned, the sophistication of technology nowadays should be able to keep ahead of criminals, as it does with debit and credit cards.

Q. What will happen if people run out of their ration before the end of the year? Will they end up cold and trapped in their homes, unable to travel?
A. The carbon ration will effectively act as a parallel currency to real money. People will learn to budget with it much as they do with money. There will not be a time when a surplus will be unavailable on the open market, though it will be progressively costly to purchase. For this reason, people will be discouraged from using up their ration and encouraged to be prudent in the same way that they are now, to avoid buying goods and services in excess of what they can afford owing to the high interest rates on debt that would otherwise have to be paid. To ease the introduction of the carbon ration, the utilities could record their customers' carbon dioxide emissions on their energy bills, as electricity suppliers are being obliged to do under new EU legislation.

Q. What about the travel requirements of politicians or people in business and the professions who claim that they have essential needs to travel long distances for meetings and conferences?
A. As with the extenuating circumstances of other groups, noted above, they will have to buy in at the market rate but in full knowledge that their ability to do so will diminish from year to year as the ration reduces and the price of any surplus rises.

Conclusions

This chapter has attempted to take into account the wishes and intentions of people in both the developed and developing world to advance their material prosperity, while focusing on the ultimate goal of ensuring a safe future for all the world's people.

The key points of the chapter can be summarized as:

- Recognition that a means must be found to ensure that total carbon savings essential to preventing serious climate change are achieved.
- Contraction and convergence is a framework representing an international solution to that end based on the principles of security and equity, with equal allowances for all.
- C&C requires that agreement is reached on just two things: the maximum

concentration of carbon dioxide permissable in the atmosphere, and the year by which there are to be equal emissions allowances for all worldwide.

- Initially, developed nations will have to reduce their emissions per capita while those in developing countries can increase theirs. However, everyone is involved explicitly in meeting the imperative of reducing carbon dioxide emissions. In the longer term, all will have a reducing ration to keep atmospheric concentrations under the agreed level.

- Carbon rationing would be a fair and guaranteed way of making carbon savings in the UK. It could be introduced now to meet the existing targets for carbon reduction of 20 per cent by 2010 and 60 per cent by 2050.

- The allocation of a personal ration would be based on the C&C principles of sharing rights to carbon dioxide emissions at the national level for the half of energy used directly by individuals.

- Trading on the open market would be encouraged, with the added benefit of administrative costs being diminished.

- The main alternative to carbon rationing is taxation. However, rationing would be easier to introduce, fairer and give a guaranteed outcome. It would promote the adoption of carbon saving as a collective goal to a greater extent than could taxation and would automatically encourage activities which are not dependent on the use of energy and, where they are, promote use of activities with lower or zero emissions.

- Promoting a 'conserver gains' rather than a 'polluter pays' strategy would have the effect of rewarding those who pursue lifestyles that benefit society as well as leading to a narrowing of the gap between rich and poor.

- Concerns about the concept of contraction and convergence and the introduction of carbon rationing, while having some justification, do not validate the case either for opposing their adoption or claiming that they are unrealistic.

Everyone involved, directly or indirectly, in promoting or engaging in energy-intensive lifestyles is inextricably drawn into playing down or avoiding discussion of the significance of the evidence of their contribution to climate change. That clearly cannot continue. Either another realistic solution to the problem of climate change must be found – and as a matter of urgency – or the C&C concept and carbon rationing must be accepted. The next chapter discusses in further detail the application of carbon rationing to individuals' lives.

8. Watching Your Figure
How to Live Within the Carbon Ration

This chapter looks at the implications for personal lifestyles of the carbon-rationing scheme proposed in the previous chapter. First, it provides the information needed to assess personal carbon dioxide emissions from household energy and travel. It then compares current average consumption with the reductions needed to prevent undue climate change. The timescale and degree of carbon reductions are discussed. Finally, guidance on how to reduce emissions is given and further sources of advice and information listed.

The most important part of the chapter concerns the data provided to enable you to calculate your own carbon dioxide emissions. For most readers, this will be an unfamiliar process. However, it is a vital starting point. A 'carbon literate' society will help people to live within a fair carbon budget determined from the planet's means to absorb a safe level of greenhouse gas emissions. In concept, this is no different from the familiar idea of a household budget in which we manage our expenditure so that we do not run into debt. We must now learn how to adjust our lifestyles so that we limit considerably the extent of the energy-dependent aspects of our lives – at home, at work, in our travel and in our leisure activities. Carbon literacy is the first step towards adopting energy-thrifty lifestyles.

Calculating your own carbon dioxide emissions

Calculating personal carbon dioxide emissions is important both to increase individual awareness and action, and as a means of enhancing collective awareness of our responsibility for climate change. At present, there is no widely available method or technique for determining personal emissions (although there are a couple of simple methods published on the internet, mentioned in 'Further Reading' at the end of the book). This chapter supplies a simple, user-friendly inventory of the carbon dioxide emissions from the principal elements of everyday living, thereby enabling calculation of personal emissions. Doing this is a necessary first step to allow people to consider how best to simplify and reduce their energy-intensive activities. Without this, individuals cannot know to what extent their current lifestyle is adding to the problems of climate change, let alone whether it is exceeding their per capita 'ration'. Just as with personal finance, where the first step to taking control is understanding where money is being spent, concern about personal carbon dioxide emissions must start with knowing what these are and where they arise.

Determining personal emissions from energy-dependent activities is likely to be salutary in many respects. Even with the limited data shown, it quickly becomes apparent from applying the figures set out in Table 4 (see below) that nearly everyone exceeds the proposed future annual ration. The benefit of the exercise will be seen in the effect it has in promoting a much wider understanding of how individuals and households use energy. Its implications are considerable for the exercise exposes fairly reliably the extent of a problem that cannot be ignored simply because it appears to be so challenging. It will also reveal in relation to overall emissions from our personal travel how much less important is the relative fuel efficiency of the mode used than is the distance we cover by that mode of transport.

A note on measurement
In this chapter alone, carbon dioxide emissions are measured in terms of tonnes or kilograms of carbon dioxide as carbon dioxide – tCO_2 or $kgCO_2$. In all other chapters, measurement is in terms of tonnes of carbon dioxide as carbon – tC. This is to follow the convention on existing websites and books, where figures to help individuals calculate their carbon emissions are reported in tCO_2. However, government statistics are published in tC, which is why tC is used elsewhere. One tonne of carbon is equal to $3.67\ tCO_2$. When reading about carbon dioxide emissions, it is always important to be certain which unit is being used.

Household energy use

The majority of the energy used in households is gas and electricity. To calculate annual emissions you need to know how much of them your household uses each year. If you have a year's worth of energy bills, you should be able to get a fair idea of this – even if some of your readings are estimated. If you have had several estimated readings, it is worth checking your meters to see how close the estimates are to reality. Your electricity bill may be separated into different tariff categories, such as off-peak, on-peak, white meter, but it is the total figure that you need. Note that gas use can be measured in several different units at the meter, including BTUs (British thermal units), cubic metres (m^3) and kilowatt hours (kWh). Bills are presented in kWh, and it is this figure you need for your carbon calculation.

Electricity: Multiply your annual consumption in kWh by 0.45 to establish the carbon dioxide emissions from this source in kilograms of CO_2.

Table 4: Household and individual carbon budgeting

Annual carbon dioxide emissions (kgCO$_2$)

ENERGY USE	Kilograms co-efficient	average household	average individual	YOU
In the household				
for each kilowatt hour				
electricity[1]	× 0.45	2,000	870	
gas	× 0.19	3,400	1,480	
for each litre				
heating oil	× 2.975			
In travel				
for each kilometre				
petrol car: as driver	× 0.20	} 2,420	} 1,050	
diesel car: as driver	× 0.14			
rail: InterCity	× 0.11			
other services	× 0.16	} 200	} 90	
Underground	× 0.07			
bus: London	× 0.09			
outside London	× 0.17	} 230	} 100	
express coach	× 0.08			
bicycle	× 0.00			
walking	× 0.00	} 0		
air[2]: within Europe	× 0.51	} 4,210	} 1,830	
outside Europe	× 0.32			
TOTAL kilograms CO$_2$		12,460	5,420	
tonnes CO$_2$		12.5	5.4	

[1] The calculation of the carbon dioxide emission co-efficient is based on the current fuel mix for electricity generation.

[2] Although varying by region and by altitude, these carbon dioxide emissions have been multiplied by an average factor of three to take account of the warming effect equivalent of other greenhouse gases in the upper atmosphere.

Sources: Various, including the Carbon Trust, IEA; DfT, National Travel Survey; DEFRA, *Environmental reporting*, 2003; National Energy Foundation; and data from Transwatch Company Limited

Gas: Multiply your annual consumption in kWh by 0.19 to establish the carbon dioxide emissions from this source in kilograms of CO_2.

Notice that current emissions per kilowatt hour from electricity are over twice those from gas use, for reasons explained in Chapter 3. (If you have signed up for a renewable electricity tariff, then of course your emissions figure for electricity will be zero.) Each of the totals from gas and electricity use should now be divided by the number of persons in the household, with the resulting two figures for the individual inserted in the right-hand column of Table 4.

Travel

Travel can be estimated separately for each method of transport. You need to estimate your annual distance travelled in kilometres for each one, and then to multiply this by the co-efficient in kilograms of carbon dioxide per kilometre ($kgCO_2$/km) in Table 4 appropriate to that method. (To convert from miles to kilometres, multiply distance in miles by 1.6.) It is unlikely you will know exactly how far you have travelled during the year, so here are some hints on how to estimate your total. Table 5 (see below) includes distances for some inter-city journeys by road, rail and air.

- **Cars:** For cars more than three years old, your two most recent annual MOT certificates, which include the total distance covered by your car, will enable you to determine your average mileage for the year. For newer cars, you could divide the total mileage to date by the age of the car in years and parts of a year to get an approximate annual total.
- **Buses:** Think about how many times a week you catch the bus, and your typical journey length. Try to include all longer-distance bus and coach journeys as well.
- **Rail:** Include regular commuting journeys, longer-distance leisure travel, and use of any underground systems.
- **Air:** For most people, air travel is infrequent enough to be memorable. Add together all your journeys during the last year, if any, both within the UK and abroad.

The aim is to get a reasonable estimate of annual travel rather than a precise figure. You are more likely to under- than overestimate because some journeys are easily forgotten – so err on the high side when making 'guesstimates'.

Emissions for car travel are given for the driver only in order to simplify the calculation and because fuel consumption does not rise very much when cars carry passengers. For passenger travel by public transport, average figures are used, covering the relatively energy-inefficient times of the many off-peak hours when there are few passengers as well as the peak hours when travel by public transport services is of course much more economical. The average difference in carbon dioxide emissions per passenger kilometre between travelling by public transport and by car is less than might be expected. For example, travelling by bus outside London (where buses have lower occupancy rates) generates very similar emissions as travelling alone by car. InterCity trains generate around half the emissions of a petrol car per kilometre, but emissions per kilometre by local trains are between those of diesel and petrol cars. The lowest are for underground rail systems owing to their relatively high passenger occupancy levels.

Table 5 (see below) records the carbon dioxide emissions for a return journey by rail for a range of UK destinations and by air for a range of overseas destinations. While one round trip by rail does not account for a significant level of emissions, the more distant destinations, particularly if regularly visited, would do so. Some examples are provided for typical journeys on the London Underground. Air travel is especially important given not only its high emissions of carbon dioxide and other greenhouse gases but also its association with long-distance travel. Each kilometre covered by air within Europe accounts for around two and a half times the equivalent carbon dioxide emissions of the same distance by car (and around five times those by train). A round flight from London to New York for a single person accounts for the equivalent of about 4 tCO_2. This is the same as the average *annual* emissions from each person in the current world's population for all their fossil-fuel needs. Against these figures, the future of air travel under carbon rationing must be seen to be particularly problematical.

Adding it all up

Once you have added up all your major sources of personal carbon dioxide emissions in Table 4, you will know your annual emissions from direct energy use. Compare this with the current individual average of 5.4 tCO_2 to see how you are doing. However, to understand your full carbon impact, it is important to remember that about half the energy in the UK economy is used by the industrial, commercial, agricultural and public sectors to create goods and services for individuals. So, your total from Table 4 should be doubled to cover your share of these non-domestic sectors of fuel consumption.

The extent of personal carbon dioxide emissions from energy use in the home is critically affected by the number of people in the household. Some-

Table 5: Carbon dioxide emissions from one return journey from London by rail to various UK destinations, by London Underground, and by air to various destinations

Destination	Round trip (km)	Carbon dioxide emissions[1] (kgCO$_2$)
Long-distance journeys by rail		
Birmingham	386	42
Brighton	190	21
Cardiff	498	55
Edinburgh	1,330	146
Exeter	664	73
Glasgow	1,334	147
Manchester	656	72
Newcastle	920	101
Reading	128	14
York	682	75
Commuter journeys by rail		
Chelmsford	100	16
Guildford	90	14
Oxford	170	27
Tonbridge	90	14
London Underground journeys		
Ealing Broadway / Victoria	16	1.1
Edgware / Green Park	12	0.8
Mile End / Holborn	11	0.8
Stratford / Marble Arch	6	0.4
Air to and from:		
Athens	4,770	1,880
Cape Town	19,100	6,410
Hong Kong	18,980	6,410
Los Angeles	17,410	6,000
Madrid	2,520	1,170
Melbourne	34,020	11,090
Moscow	4,990	1,910
New York	11,070	3,930
Paris	690	560
Rome	3,140	1,340
Tel Aviv	7,200	2,660
Tokyo	19,300	6,470

[1] The calculations for carbon dioxide emissions from air travel have been multiplied by three to take account of the warming effect equivalent of other greenhouse gases in the upper atmosphere, and are based on typical seat occupancy rates of 80 per cent.

Sources: CfIT, *A comparative study of the environmental effects of rail and short haul air travel, 2001*; IPCC, *Special report on aviation and the upper atmosphere, 1999*

body in a one-person household, regardless of income, uses around twice as much electricity and gas and therefore produces twice the emissions as somebody in a three-person household. Thus people in one-person households have to do considerably more to lower their emissions than a person with the same lifestyle living in a multi-person household.

Making awareness of our carbon dioxide emissions part of everyday life

Introducing carbon rationing would require information on the carbon dioxide emissions inherent in items purchased and used, and in modes of transport, to be as readily available as price. Many changes will be required to achieve this for consumers. With carbon rationing, carbon becomes a parallel currency. Therefore it will be essential that, as a minimum, 'carbon information' is provided in the same way that price information is now presented. There are a couple of moves in this direction. Emission figures (in terms of grams of carbon dioxide per kilometre – gCO_2/km) are published in advertising material for new cars. In addition, there are plans at an EU level to include emission figures on energy bills, but it could take a long time for these to become legislation.

The following would improve householders' understanding of the carbon dioxide emissions that result from each activity so that they would be able to relate these easily to their annual ration:

- **Smart bills:** Including carbon dioxide emissions for gas, electricity, fuel oil and other fuel bills.
- **Smart meters:** Gas and electricity meters to be upgraded over time to include a running total of carbon dioxide emissions, and provide comparisons with previous periods.
- **Smart receipts:** Including emissions on petrol and diesel receipts.
- **Enhanced petrol pumps:** Displaying emissions as well as price and quantity.
- **'Carbon-ometers':** Adding a carbon counter to standard car, motorcycle and moped displays, allowing the driver to have a record of total emissions, plus a trip carbon calculator (equivalent to the mileometer).
- **'Carbon responsibility' in advertising:** All flight tickets and travel promotional material (such as adverts in the media, outdoors and on the internet) to include equivalent carbon dioxide emissions.
- **Carbon labels:** Energy labels on appliances and light bulbs to include average annual emissions – or a range of figures if preferable.
- **'Carbon promises':** Insulation materials (such as loft insulation) and home

improvements (such as double or triple glazing) to be promoted in terms of the carbon as well as energy savings they can provide.

- **Carbon-rated homes:** All houses, new and second-hand, to be sold with an energy survey and an estimate of average annual carbon dioxide emissions, plus tailored advice on how to reduce the emissions.

- **'CarbonWatchers':** A community information and support scheme equivalent to diet schemes such as WeightWatchers. Based on the diet clubs template, it would provide its members with booklets explaining the 'carbon impact' of different purchases and travel options, set reduction targets for individuals, hold regular audits (the equivalent of weigh-ins) and provide both professional and peer support for participants. Monitoring personal emissions will provide a practical means of appreciating the role of different energy-dependent activities in the total 'carbon budget' and of finding ways to avoid excess consumption.

With proper information and advice, people should find it possible to be able to use this new 'currency' quickly, particularly as it will become an increasingly important part of life. Prior to the recent changeover to the euro in many European countries, fears were expressed that many people, particularly the vulnerable, would find it hard to adapt from their familiar national currency. In the event, very few problems were reported. As money is important, it was in people's self-interest to come to terms quickly with the new currency. Also, governments provided a lot of information in imaginative and effective ways. In the Republic of Ireland, for instance, everyone received a set of replica cardboard euro coins and notes before the real ones were available to increase early familiarity with the new currency. Similarly, people could be helped to understand both their carbon ration and the new carbon currency for energy by employing the methods suggested above.

This book is suggesting that householders initially be made responsible only for their direct energy usage. However, it is clear that individual choices about what sort of products are bought also has a knock-on effect on the fuel used in the industrial and commercial sectors. As the section on food in Chapter 3 demonstrated, food can have a very different 'indirect' energy content, depending on how and where it was grown, transported from and processed. Carbon rationing could be extended to products which could be labelled to show their indirect carbon dioxide emissions. However, as discussed earlier, doing this could overload consumers with information as well as being difficult to provide, and is not recommended for the initial years of carbon rationing. As a version will also be in place for businesses, the prices of carbon-intensive products would increase, automatically directing consumers towards the cheaper items whose production entails fewer emissions.

Future rations

This section illustrates what future personal carbon rations are likely to be under two different scenarios, and compares this with current average emissions in the UK. The difference between today's consumption levels and those that will be required just ten or twenty years into the future is striking and leaves little room for complacency.

The level of future rations depends on what cuts need to be made to ensure that the agreed level of carbon dioxide emissions in the atmosphere is not exceeded. It also depends to a lesser extent on the date chosen for global convergence on equal emission rights. This book has argued that a concentration of 450 ppm carbon dioxide in the atmosphere is the maximum which should be considered safe and indeed that, given the uncertain response of the planet to additional climate change, 450 ppm may turn out to be too high. Rations have been calculated both on the basis of the government's 60 per cent reduction target for 2050 (designed to stabilize concentrations at 550 ppm), and on the basis of the reductions that would be needed to stabilize at 450 ppm. As with global carbon reductions based on contraction and convergence, rationing would be introduced in a phased way. It would be totally unrealistic to aim for an immediate reduction to the level recommended to avoid serious damage to the planet's climate. We suggest that the principle of convergence must be built into the framework to achieve the reduction, that the target date must be 2030, and that the concentration of carbon dioxide in the atmosphere must not exceed 450 ppm.

Table 6 (see below) shows both the total per capita ration (including industrial and commercial energy use) and the personal carbon ration figure from 2005 to 2050 under two scenarios for carbon reduction. Rationing is assumed to begin in 2005, with the initial ration set at the expected average for that year, and then decreasing yearly after that date.

Compared with expected average emissions in 2005, the 550 ppm scenario requires a personal reduction of 63 per cent by 2050, and the 450 ppm scenario an 80 per cent reduction by 2050. In both scenarios, the ration shown will be equal for everyone in the world in 2050, and in the 450 ppm scenario it will be equal by 2030. Remember from Chapter 2 that stabilizing carbon dioxide concentrations – at any level – requires the eventual reduction of global carbon dioxide emissions to a fraction of their current levels. So, under both scenarios carbon rations will have to continue to fall from 2050 into the future.

Table 6: Estimated carbon rations from 2005 to 2050 under different reduction scenarios. Rations are shown for all energy, with personal energy use in brackets (50 per cent of total)

Current and projected average carbon emissions per person tCO_2 (includes carbon equivalent emissions from aircraft)	Future carbon rations, per person tCO_2 All energy use (Personal energy use)	
	Government proposal: 550 ppm, convergence at 2050	Our proposal: 450 ppm, convergence at 2030
2002 10.7 (5.4)		
2005 10.4 (5.2)	10.4 (5.2)	10.4 (5.2)
2010 10.8 (5.4)	9.6 (4.8)	8.9 (4.5)
2020 11.7 (5.9)	8.2 (4.1)	6.0 (3.0)
2030 No CO_2 projections are	6.8 (3.4)	3.0 (1.5)
2040 available after 2020	5.3 (2.7)	2.6 (1.3)
2050	3.9 (2.0)	2.1 (1.1)

Note: The projections for 2005 to 2020 are based on government figures.

The figures in the table should be treated as approximate rather than precise. They rely on linear reductions (the same annual fall in emissions) and current population projections. The 550 ppm ration figures are based on those of the RCEP, the 450 ppm ration figures are based on calculations by the GCI. Under the 450 ppm scenario, just one return flight from London to Athens would exceed the whole personal carbon ration for the year in 2030. In the 550 ppm scenario, one return flight from London to New York would exceed the ration by 2030.

These figures should shock you – a state which will be further reinforced by comparing your total with the graded reduction needed in the transition to an equal per capita level as soon as possible. As Table 6 shows, by 2030, under the government's proposal, the annual individual ration has to be reduced to 3.4 tonnes, while under our proposal, based on what we believe to be the reduction genuinely needed, it has to be down to 1.5 tonnes. You should now appreciate the scale of the challenge.

However, do not despair. Energy-use patterns have changed considerably over recent decades, with energy used for personal travel almost doubling since 1970. Under the 450 ppm scenario, carbon dioxide emissions from personal travel would have to halve over the next twenty years. If energy efficiency and low-carbon technologies are used in parallel with a significant reduction in motorized travel, this will not represent a much greater rate of change in mobility than the UK has already experienced – it has just been in the other direction. This degree of change is not necessarily going to be easy, but it is not unrealistic.

The figures in Table 6 also assume that adults and children get equal personal carbon rations. As mentioned briefly earlier, the rationing scheme would have to be fair to children, but not unfairly advantage people with larger families. This would probably mean that children would receive somewhat less than the adult ration. The children's ration would be based on evidence of the emissions currently generated in households with and without children.

How you can reduce your carbon dioxide emissions

Climate change cannot be limited solely by the actions of individuals. Throughout this book, the emphasis has been on the need for the world community to act together as the only possible solution to limiting greenhouse gas emissions. However, individuals can reduce their own carbon dioxide emissions and this advice is meant to help you reduce your 'carbon impact' on the environment and thereby make a contribution. By demonstrating that a low-carbon lifestyle is possible, you will also help in the process of persuading the wider public and the government that carbon rationing can work. As with the country as a whole, you should first cut down on energy-related activity, secondly look at efficiency options and then at renewable sources of energy. But, to reiterate, at least as important as 'doing your bit' is helping society as a whole move towards a global agreement on a C&C basis and for the UK to adopt carbon rationing.

This chapter has already emphasized the importance of individuals auditing their own annual carbon dioxide emissions as a starting point, and has provided a tool for doing so in Table 4. However, once you know approximately what your emissions are, what should you do next? This section provides some suggestions for action and identifies other good sources of information about how to reduce personal travel, particularly by car, and how to save energy in the home.

Personal travel

The key advice for personal travel that uses energy is to travel less. This is particularly true for air travel, but also for travel by car and public transport. That is not to say that cutting down will be straightforward: it may need a lot of thought and planning and require years to make the necessary lifestyle changes. Beyond cutting down on travel, a switch can be made to modes of transport with lower carbon dioxide emissions (see Table 4), to sharing lifts or buying a more efficient car. Taking up cycling is, for many people, a good option for journeys of up to 4–6 kilometres. To do so saves money and improves health. The organizations and websites listed below can provide much more detailed information, including local options for avoiding or cutting down on car use.

Step 1: Understanding your own travel energy use

Write your own transport diary for a week or a month (and get everyone else in the home to do the same). Include walking and cycling. You should note down: day of the week, time, origin, destination, aim of the journey, method of transport, cost and trip time. The results will give you a better insight into how and why you make each journey – particularly motorized journeys. This information will be critical to help you prioritize changes in your patterns of travel.

Step 2: Changing your travel patterns

Although transport for short journeys is most easily replaced by non-motorized methods, remember that in the end distance is the crucial determinant of your carbon dioxide emissions. You may be able to achieve more by cutting out one long-distance weekend journey than months of avoiding using the car to visit the local shops.

Changing your travel patterns could include:

- Walking and cycling for local trips (see 'Sources of information' below for information and advice).
- Finding out about local public transport and using it instead of your car.
- Combining several purposes in one journey, which will also save your time.
- Joining a local car-share scheme – these are available in some areas of the country.
- Using more local shops and other services to save longer-distance travel.
- Taking holidays closer to home.
- If you have a car, setting yourself an achievable reduction target for the year (say, 5–10 per cent or 1,000–1,500 km) and check whether you are meeting

the target during and at the end of the year. Do the same thing the following year.

Your travel diary (see Step 1) will give you the best idea on how to cut down on motorized travel.

Step 3: Reducing carbon dioxide emissions from your car
The government gives the following advice for reducing emissions (and saving money) when using your car:

- **Plan ahead:** Choose uncongested routes, combine trips, car share.
- **Cold starts:** Drive off as soon as possible after starting.
- **Drive smoothly and efficiently:** Harsh acceleration and heavy braking have a very significant effect on fuel consumption; driving more smoothly saves fuel.
- **Slow down:** Driving at high speeds significantly increases fuel consumption. Driving at 70 mph uses 30 per cent more energy, for instance, than driving at 50 mph.
- **Use higher gears** as soon as traffic conditions allow.
- **Switch off:** Sitting stationary is zero miles per gallon; switch off the engine whenever it is safe to do so.
- **Lose weight:** Don't carry unnecessary weight, remove roof racks when not in use.
- **Regular servicing** helps keep the engine at best efficiency.
- **Keep the pressure up:** Make sure the tyres are inflated to the correct pressure for the vehicle.
- **Do not compromise** on safety but be aware that the use of onboard electrical devices increases fuel consumption.
- **Check your fuel consumption:** It will help you get the most from the car; changes in overall fuel consumption may indicate a fault.
- **Use air conditioning sparingly:** Running air conditioning continuously will increase fuel consumption. If it's too hot to drive without air conditioning, ask yourself if you really want to spent that time on the road, rather than somewhere more congenial where you can enjoy the sunshine.

Sources of information: car travel and alternatives
Local councils generally have initiatives on environmentally friendly forms of travel. They will be able to give advice on alternatives to the car in your area. Often there will be a dedicated 'green travel' adviser. Most also have websites which may contain useful information. The website addresses are in the format: www.*yourtown*.gov.uk or www.*yourcounty*.gov.uk.

There is a very useful little book that includes a lot of helpful hints and tips, case studies and success stories, together with tables helping you to work out how much money you could save by using your car less or giving it up altogether. It also provides references to further information: A. Semlyen, *Cutting your car use: save money, be healthy, be green!*, revised and updated edition, Green Books, Totnes, 2003 (tel. 01803 863260).

Sustrans is a sustainable transport charity which is best known for its work on the National Cycle Network. Its website offers information and links to free local cycle maps, as well as maps for sale: www.sustrans.co.uk; Sustrans information line: 0117 929 0888.

There are also other cycling organizations which can advise on taking up cycling, for example the UK's national cycle organization, CTC (www.ctc.org.uk, tel. 0870 873 0060), or local organizations such as the London Cycling Campaign (www.lcc.org.uk, tel. 020 7928 7220).

Sources of information: air travel and alternatives

As this chapter has clearly proved, air travel is by far the most polluting form of travel. The following websites focus on the effects of air travel and alternative methods of travelling long distances.

'Flying off to a warmer climate' is a site which explains the climate change issues concerning air travel, and allows calculations to be made of emissions from particular flights. It also has some suggestions for alternative methods of transport: www.chooseclimate.org.

The quirkily named website of 'The man in seat sixty-one' is an independent and comprehensive guide to travelling by train and ship both within the UK and Ireland, and to the rest of Europe and beyond. It is an extremely helpful and well-designed source of information, with links to companies from whom tickets can be bought: www.seat61.com.

It is possible to buy carbon 'offsets' for air travel (and other energy-using activities), where companies guarantee saving or re-absorbing from the atmosphere the same amount of carbon dioxide as a passenger's flight will emit. While this goes some way to acknowledging the problem, as Chapter 6 has shown, carbon offsets are expected to have a limited role in reducing emissions. There are not enough offsets available to allow everyone to keep on flying. Perhaps the most positive use of carbon offsets is to invest in them in parallel to reducing your energy use, rather than using them to 'justify' a high-energy lifestyle.

Household energy use

There are many ways of reducing household energy. Unless you live in an extremely efficient house (which probably applies only to a few thousand people in the UK), there are considerable opportunities for saving energy, many of which will not require major lifestyle changes or financial investment. Key examples are given here, but the information sources listed below will be able to give more comprehensive and tailored advice.

Step 1: Understand your current energy consumption

Having looked at your annual energy consumption in order to audit your carbon dioxide emissions, it is worthwhile considering in more detail how that energy is used, so you can identify the major opportunities for saving.

The split of energy use between heating and hot water depends very much on your house and style of life. For gas central heating, the average split of energy use has been estimated as: 70 per cent space heating; 28 per cent water heating (and 2 per cent for cooking with gas). This split between heating and hot water also applies to other fuels. A more efficient or newer house will use less heating energy; large, inefficient or old homes will use more heating energy; households with more people will use more hot water. Think about your own household and how you might differ from the average.

How electricity is used in your home will again depend on what lights and appliances you have and how you use them. The average for the UK is shown in Table 7 (see below). Notice that the two top categories are fridges and freezers, and lighting. Lighting can easily and cheaply be made more efficient, but the same is not true of fridges and freezers.

Step 2: Free and cheap options

1. Learn to use your heating and hot-water system as economically as possible. There is a long history of research which shows that many people do not really understand the controls that come with their heating and hot-water systems. So if you fall into this category, don't worry, you are not alone! However, learning to control your system should allow you to save energy (and money). It can also be very worthwhile to have controls fitted if they are missing – speak to some of the professionals identified in 'Sources of information' below.

This book can't give detailed advice for all heating systems, but gas and other non-electric central heating systems may have the following elements:

• **Timer:** This allows you to control when your heating and hot-water switches on. To save energy, experiment with turning the heating off before you go

Table 7: How electricity is used within the average home in lights and appliances

End use	Percentage of electricity for lights and appliances
Cold (fridge, freezer, fridge-freezer)	24
Lighting	24
Cooking (hob, oven, kettle, plus other)	17
Wet (washing machine, tumble dryer, dishwasher)	16
Consumer electronics (TV, video, satellite, PC, etc.)	14
Miscellaneous	4

to bed – the better insulated your house is the longer it will take to cool down, but it should stay warm enough for at least thirty minutes. Switching off heating half an hour earlier could save over 5 per cent of your energy bill.

- **Hot-water thermostat:** This controls the temperature of stored hot water and should not be set higher than 60°C (both to prevent scalding and save energy).
- **Room thermostat:** This sets the temperature for all your rooms. As a rule of thumb, for each 1°C you reduce the thermostat, you will save 10 per cent of heating energy. You will have to make your own judgement on the balance between energy saving, wearing warm jumpers and personal comfort. (Note that living-room temperatures below 16°C are considered prejudicial to health; 18°C is the normally recommended minimum.)
- **Thermostatic radiator valves:** These allow proper control over radiator temperatures and are particularly useful for setting lower temperatures in less-used rooms, or bedrooms, which most people prefer to have cooler than their living areas.

Gas fires are generally considerably less efficient than central heating systems. This is particularly true of some open flame and decorative fires, which can deliver as little as a quarter of the energy used as heat to the room (the rest goes up the chimney). Don't turn down the central heating radiator in your main living room and turn up the gas fire – do it the other way round and save energy.

2. Bathing and showering options. Energy use for hot water when bathing and showering depends on the temperature of the water and how much hot water you use and how often. These are all things you can control.

In a comparison between baths and five-minute showers, perhaps surprisingly the power shower uses most water, at 100 litres, the bath is next, at 80 litres, then the standard shower at 50 litres, an aerating shower (which saves water and energy by adding air to the mix) uses 25 litres, and an electric shower just 15 litres. As the average time spent in the shower is nearer 7–8 minutes, in fact only aerating and electric showers use less water than a bath. Personal hot-water usage has risen considerably over recent decades for complex and intertwined social and technological reasons – the rise of the daily shower is one factor because the switch to daily showering from less frequent bathing has generally increased consumption of water and energy.

If you have a standard or power shower, you can probably fit a water-saving shower head or a flow restrictor to save water, money and energy while still experiencing a good-quality shower. However, the non-technological options are more powerful. To state the obvious, you can save energy by having cooler, quicker or less frequent showers and baths. For example, shaving just two minutes off your showering time could save a quarter of the energy used.

3. Savings on lights and appliances electricity use. Free and cheap options in approximate order of priority include:

- **Lighting:** Installing energy-saving light bulbs (technically known as compact fluorescent lamps or CFLs) is the best option. Research shows about half the light fittings in the average house can take these without adjustment. Installing CFLs in the four most used lights could save 200 kWh per year, more than a quarter of electricity typically used for lighting. But don't stop at installing four – you can install many more than this. Energy-saving bulbs generally cost less than £5 (and can be very much cheaper) and come in a reasonable range of sizes and designs. They will save their higher cost several times over as they use much less electricity and they last much longer (usually ten times) than conventional bulbs. As with all bulbs, switch them off when they are not needed. If you don't like the whiter light that some energy-saving bulbs emit (compared with traditional bulbs), use them with a creamy or yellow-coloured lampshade, to change the colour of the light.
- **Saving on standby:** Turning the television and other gadgets and entertainment equipment off standby (you may need to unplug them to do this) can save up to 10 per cent of your electricity. Equipment chargers that are plugged in, even when nothing is attached, are using energy (feel them – they're warm; that's electricity being wasted and turned into heat), so switch them off at the plug when not in use.

- **Washing machines:** 60°C and 40°C are already the most popular washing temperatures. It is worth making the switch from 60°C to 40°C as this can save 40 per cent of energy per cycle. Most people underload their washing machines. Washing machines are made for loads of 4.5–5 kg, so see if you can squeeze a bit more in and cut down the number of cycles. Also, if your hot-water supply is heated by gas, supplying 'hot fill' water to the washing machine will save on carbon dioxide emissions, compared with the washing machine using its own electric heater to heat the water.

- **Dishwashers:** As with washing machines, lower-temperature cycles save energy (a 55°C cycle uses around a third less energy than a 65°C cycle), as does properly loading your dishwasher and so reducing the number of times it is run. Washing up by hand *probably* produces lower emissions than using a dishwasher, but the evidence is not strong enough to be certain.

- **Kettles:** Kettles use a surprising amount of energy – typically around two-thirds as much as an oven or hob during the year. The rather obvious energy-saving tips are to boil only as much water as you need, and not to re-boil (because you walked away from the kettle before it was ready and then had to boil it again). Using a gas hob-top kettle can save more than half the carbon dioxide emissions of an electric kettle.

- **Taming your microwave:** If you have a microwave with an electronic clock display (as is the case for most microwaves), powering the clock could be using as much energy per year as is used for cooking. Switch off the microwave completely when not in use.

- **Cooking:** Using your microwave instead of a conventional oven will save energy. Other tips are putting the lids on pans of boiling water, not preheating the oven in advance of when you need it, and, for electric hobs, using pots and pans which are not warped and have good contact with the surface.

- **Fridges and freezers:** Their energy consumption is not hugely influenced by householder behaviour, but all the following will save energy: defrosting the freezer, removing dust from the cooling coils at the back of the appliances (in some newer appliances these are enclosed, so this does not apply), not putting hot food into the fridge.

Step 3: Options requiring more investment

The most effective investments in energy saving (both in terms of energy saved and pay-back on investment) depend on your individual property and circumstances. The following options are usually the most worthwhile:

- Fitting additional loft insulation – most houses now have some loft insulation, but generally far less than today's standard. This is one of the most cost-effective ways of saving energy. You can fit it yourself, or have it done by a professional, and your energy company / local council may offer subsidies or discounts.

• If you have unfilled cavity walls, you should nearly always get them insulated. This could save up to 30 per cent of your heating energy and will pay back your investment in just a few years.

To cut down on heat loss, other home improvements that will save energy include: draught proofing (being careful to ensure adequate ventilation remains); fitting shelves above radiators under windows; fixing reflective foil behind radiators on external walls; lagging hot-water pipes; adding thermal linings to your curtains.

If you are interested in installing your own renewable energy source, probably the most cost-effective option will be solar water heating. There are several companies which can fit these, and in some areas of the country there is support for people who want to learn to fit their own solar water heater. However, as with any purchase, you need to be wary of less reputable companies who make exaggerated claims – the most you are likely to save is half your hot-water bill, which amounts to approximately 15 per cent of your gas bill per year. Fitting solar hot water, for most people, is a statement of conscience and an investment in the future, rather than a great financial investment.

Other, more expensive, options include solar PV and geothermal energy. There are some grants and subsidies available for all these technologies, which could help offset the cost.

Finally, burning wood to heat your home is another renewable energy option. However, there are restrictions on the use of non-smokeless fuels in built-up areas (which in some cases can be complied with by investing in a wood burner which guarantees clean emissions). Burning wood requires considerable storage space, investment in a wood-burning stove, and in a poorly insulated home will not provide a large proportion of heating energy. In a ranking of options based on carbon saved per pound invested, it is likely to come below most insulation improvements, but nevertheless is one of very few options available for renewable heating energy. (By contrast, burning coal or smokeless fuel in an open fire is probably the highest-carbon heating option you could choose, both because coal is carbon-intensive and because open fires are very inefficient.)

Step 4: What to do when buying new equipment
When replacing appliances, check for energy labels so that you buy an energy-efficient model. Energy labels are on all fridges, freezers, fridge-freezers,

washing machines, tumble dryers, dishwashers, electric ovens and light bulbs. However, remember that energy consumption is more important than energy efficiency. A large, efficient fridge-freezer may use more energy than a smaller, less efficient one. Details on total energy consumption are given on the energy label and this is the most important information. Small can be beautiful.

If you are buying a new boiler for hot water and heating, talk to your plumber about getting a condensing boiler. These are around 90 per cent efficient compared with around 80 per cent for an average new boiler, or 65 per cent for the average boiler installed at the moment, and are suitable for nearly all homes.

When replacing your windows, invest in a more efficient type. Double glazing is now standard, so consider 'low-e' glass (which saves more energy) or triple glazing, particularly for rooms kept at warmer temperatures.

Step 5. What to avoid doing

- Buying a patio heater / air conditioning / a large, frost-free fridge-freezer / power shower.
- Opening a window to cool the house (in winter!).
- Having a 300–500W security light switching on all the time outside your house. (If you need a security light, there are efficient alternatives.)
- Heating your conservatory. Amazingly, conservatories that should help warm homes in the autumn and spring are used as heated spaces in winter. This is incredibly wasteful.

Sources of information

There is a national network of over fifty Energy Efficiency Advice Centres, which provide a wealth of advice and expertise on energy efficiency that is free, impartial and locally relevant. The centres usually help you complete a free energy 'home check', on the basis of which they can evaluate the efficiency of your home and guide you towards the products which could help you make maximum savings. They will also be able to advise you if there are any grant, subsidy or cheap loan schemes in your area for which you could qualify. Call freephone 0800 512 012 to be put through to your local energy efficiency adviser.

As for travel, local councils are often a good source of advice to help you save energy at home. Some local councils also have discount schemes for fitting solar water heating, purchasing efficient boilers, buying insulation material and other energy- and carbon-saving options.

The national organization that provides advice and information on energy savings is the Energy Saving Trust. The 'save energy' part of their website contains a lot of very useful energy-saving ideas, a database of the grants available and lists of energy efficient products: www.saveenergy.co.uk; energy efficiency helpline: 0845 727 7200 (general website: www.est.co.uk).

Your energy company also has an obligation to supply you with advice on how to save energy. You can phone and ask for their energy-savings brochure. Some also offer a free home energy survey. In order to help meet required energy-savings targets, companies usually provide financial incentives to customers to take up energy-efficiency measures, typically cavity wall insulation, top-up loft insulation and energy-saving light bulbs. Contact your company to find out how it can help you save energy and reduce your carbon dioxide emissions.

It is now possible to buy your electricity on a renewable energy tariff. Although, as Chapter 6 explained, renewable energy by itself cannot solve all our climate change problems, supporting renewable energy is a step forwards as long as signing up to the tariff is not seen as an excuse for not making every effort to reduce your consumption. GreenPrices website (among others) offers comprehensive information and price comparisons for UK renewable energy suppliers, and also includes useful discussions about the issues involved in buying renewable electricity: www.greenprices.com.

Getting inspired

Another way of getting further information and inspiration is to visit an environment centre or a sustainable housing project. Probably the best-known environmental information centre in the UK is the Centre for Alternative Technology in Machynlleth, Wales (website: www.cat.org.uk; tel. 01654 705950). This excellent eco-centre was founded in 1975 and has a wide range of different 'eco' buildings to visit, displays about environmental building, renewable energy and energy efficiency, among other environmental topics, and a good bookshop. In addition, it runs a range of short courses on the environment. There are several other environmental education centres, including the Earth Centre in Doncaster.

Alternatively, visit a local sustainable housing project, to see real people trying out different technologies and ways of life. There are projects in nearly all areas of the UK. Details are available in the following guide: N. White, *Sustainable housing schemes in the UK: a guide with details of access*, Hockerton Housing Project, Nottinghamshire, 2002. Available from Hockerton Housing Project (website: www.hockerton.demon.co.uk; tel. 01636 816902).

Non-energy options

This book has concentrated on direct uses of energy. However, as was mentioned earlier, the goods we buy come with energy 'embodied' in them: that is, the energy that was used in their manufacture, transport and retailing.

There is very little product information available to enable consumers to make decisions based on the energy inherent in the products they buy. There are some 'rules of thumb', however, about how to choose lower-energy products:

- For food and drinks, use the 'country of origin' information to choose those which were grown or produced closer to home, to reduce your 'food' miles.
- Consider buying some UK organic food, especially that which is locally grown – organic food that has been transported a long distance (as some currently is) is probably not the right choice in terms of reducing carbon dioxide emissions.
- Choose more seasonal food, which is less likely to have been grown either far afield, or in heated greenhouses in the UK.
- Find out about your local farmers' market, farm shops or vegetable box scheme, all of which can supply local, fresh food.
- Buy recycled products, or those with high recycled content – good examples include paper products made with so-called 'post-consumer' waste.
- Buy products which are recyclable or whose packaging is recyclable / re-usable (and make sure you do re-use / recycle at the end of their life).
- Buy better-quality products which are likely to have a longer life. Owning fewer but longer-lasting objects throughout your life will be less energy intensive than having to buy frequent replacements.
- Choose products which will be repairable if this is an option.
- Avoid disposable products where possible, as these usually have a higher 'embodied' energy than the non-disposable alternative. Examples include using your own shopping bag rather than hundreds of disposable ones per year, buying rechargeable batteries, using a fountain or cartridge ink pen rather than a throwaway one.
- Choose products which have the European 'Ecolabel' (a flower-like symbol). This indicates that the product has been independently assessed and found to meet strict environmental criteria (considering more than just energy consumption), putting it among the best in its class.

It must be stressed that these rules of thumb may not apply in all situations, but the general principles should help you reconsider your purchasing behaviour.

You can also reduce your 'energy impact' by reducing the amount of waste you produce, and by managing the remainder carefully. Waste is a growing problem and throughout Europe waste generation appears to be increasing at a *faster* rate than economic growth. As mentioned earlier, the management of materials and waste has a strong link to energy use, and waste embodies wasted energy as well as materials. Better management of waste can result in energy and carbon savings, and these are outlined below.

In order of priority you should:

- **Reduce the amount of waste you produce:** Probably the single best thing you can do is compost your vegetable and garden waste if you have a garden (or, if you don't, use a wormery – an easy and efficient system of converting ordinary kitchen waste into top-quality compost and concentrated liquid feed through the natural action of worms). Not only does this reduce the amount of waste for disposal by up to a third, it also provides a valuable resource for your garden. Many local councils offer advice on composting as well as discounts on compost bins. Rethinking and altering purchasing habits, as suggested above, are also key to reducing the amount of waste you produce.

- **Re-use your waste materials:** There are many options for turning what you no longer want into something useful. Some examples are: giving saleable clothes, books, toys, bric-a-brac to your local charity shop (or selling them yourself); using glass jars for homemade jams and pickles; using shoe boxes / biscuit tins for storage; putting old clothes and sheets into a childrens' dressing-up box; growing seedlings in yogurt pots.

- **Recycle your waste:** For the majority of waste materials, recycling saves energy and carbon compared with disposal and manufacturing from virgin materials. How convenient you find recycling depends on the facilities you have locally; in most places there are now door-to-door collections of recyclable material or local recycling banks. Try to use your local facilities as much as possible – remembering to avoid compromising the carbon-saving benefits of recycling by making special trips by car to deliver recyclable materials! In addition, some charities accept special items such as mobile phones or printer cartridges for recycling / re-using. Your local council is likely to have produced a recycling guide which will list many possibilities.

The advice in this section has been aimed at you in your role as an individual and a consumer. However, people are also employees, employers, students, volunteers, members of families, friends, involved in local communities, share owners, voters and citizens. In all of these roles you can have an influence on the level of carbon dioxide emissions generated by activities in which you have a stake.

Changes over time

Beyond the immediate and medium-term actions that can be taken to reduce personal carbon dioxide emissions (anything from switching the television off stand-by to replacing your car with a smaller, more efficient model when the time comes), perhaps the more difficult task is to think now about what adaptations need to be taken for a longer-term, low-carbon future.

As the previous section has illustrated, as a minimum under the government's scenario, individuals will be required to make a 60 per cent reduction in emissions by 2050. Under the reduction suggested in this book, which we believe to be more realistic, personal emissions will have to reduce by over 70 per cent by 2030. In parallel to the national picture presented in Chapter 6, although energy from increased efficiency and renewable sources can help with these reductions, considerable changes to people's lifestyles will also be required. Many of these changes should be positive, leading to better health, quieter and safer streets, and so on, but they run counter to current trends in society and require thought and commitment.

Reducing dependence on motorized travel is a long-term project. Although it may be straightforward to decide not to fly, or to fly less often for holidays, it will be more difficult to disentangle everyday activities from dependence on the car. Patterns of travel to work, to school, to visit relatives, to do the shopping and for leisure all have to be rethought. In some cases, non-motorized alternatives do not exist. If you live thirty miles from your work, for instance, cycling or walking are not possible. The question is whether in the course of time you can move closer to your job, change to a job that is closer to where you live, buy a more efficient or smaller car, start lift-sharing, use public transport (where that would generate lower carbon dioxide emissions), try and work at home for part of the week, or if you must continue to travel that distance, cut down on travelling for other, non-essential activities. Each of these responses would save a different amount of carbon and involves a different balance of costs, changes in lifestyle and improvements in quality of life. There is no single right answer. But as carbon rations decrease, and the need to save energy increases, more demanding and effective steps are likely to become increasingly necessary.

There will be opportunities for making considerable changes to your daily transport patterns when you switch jobs or move home. This may well be when the most substantial reductions can be set in train as the effect will, of course, be longer-term than any reduction that can be achieved on the type of journey that is made less frequently. Before deciding on your new location, you should calculate the different levels of carbon dioxide emissions entailed by each option (using Table 4) and use this information to help live a lower-carbon lifestyle. The option entailing the least emissions is also likely to save you travel time.

The introduction of carbon rationing will also bring in a self-reinforcing social and financial system that enables and rewards people for living within their ration. Individuals will not be left on their own to make self-sacrificing choices. Carbon literacy, information and advice will help people to live within or below their ration. Those unable to do so will have to purchase on the open market the surplus of those who can. But, with the phased strategy, it will soon become apparent that this surplus will also reduce as essential needs take up a rising proportion of the ration, year on year. In these circumstances, it will be progressively necessary and rewarding to live within the budget as any surplus will be able to make a rising contribution to income. In the process, energy-thrifty lifestyles will be steadily pursued and relentlessly promoted. Conversely, it will become increasingly expensive from year to year to buy the surplus of anyone's ration on the open market, and it is very likely that this will be at an exorbitant rate in later years. Heroic individual struggle should not be required: carbon rationing will involve societal change where everyone changes their lifestyles, and choices to reduce emissions are applauded and rewarded.

Lifestyle changes

Carbon rationing will inevitably involve changes in lifestyle – both welcome and unwelcome. Although people will be able to choose how to live within their ration, and which aspects of their lives to change, given the lack of technological fixes for motorized transport, travel options are likely to narrow. At present, travel to distant locations is seen not only as a benefit in its own right, but as symbolic of social and economic success. However, under carbon rationing there will be much less scope for this, particularly as regards flying to foreign countries. There can be little doubt that many will view the resulting prospects as a considerable limitation on their choice and a distinct reduction in their quality of life. Also, many will lose the closeness of their international networks of family and friends.

Opportunities to maintain and acquire second homes or retirement homes in distant rural locations on the Continent, such as in Spain, France and Italy, and even farther afield – let alone in the UK – have grown apace in the last few decades. It is very unlikely that people will be prepared to use such a critical proportion of their ration for this purpose, particularly if it entails flying or other long-distance travel.

Decisions to emigrate will not be so readily taken when it is realized that links with the 'mother' country can only rarely be maintained, for instance travelling to and from Australia to 'keep in touch'. This will also have consequences for the composition of the future population of the UK.

Likewise, in view of the high transport costs, the prospects are dire for activities such as international business, professional and political conferences (including ones on climate change and its implications), all international events, skiing and holiday cruises. Particularly hard hit will be sporting activities and competitions, not only for players and staff but for spectators too, such as the Olympics, cricket, rugby, golf and especially football, which is organized on the basis of regional, national and international matches. The personal 'ration' is likely to allow for only the most essential of such long-distance journeys.

There are, however, likely to be considerable indirect social benefits to carbon rationing, including the following:

- **A healthier population:** Far more journeys will be made on foot and by bicycle, leading to many more people taking exercise on a daily basis, and thereby improving their health and wellbeing.
- **Better food:** General health will improve owing to the greater availability of better-quality food as the 'carbon impact' of fresh, local, organic food tends to be lower than that of frozen, pre-prepared, chemical-laden alternatives.
- **Fewer oil spills:** Less oil transported around the globe will result in fewer accidents at sea.
- **Less danger on the roads and fewer deaths and injuries:** Significant benefits such as these will be the outcome both of less traffic and lower speeds.
- **Better resource use:** A low-carbon economy will enable other environmental problems to be solved as a result of a more efficient approach to using energy to transform resources into products.
- **Reduction in fuel poverty:** Comfort levels will be improved as the thermal condition of housing is increased to minimize use of the carbon allowance to keep warm.
- **A quieter life:** A lowering of ambient noise levels will result, owing to the reduction in road traffic, lower speeds and, for those on or near flight paths, far fewer aircraft in the skies.
- **More pleasant travel:** Lower levels of traffic and lower speeds, reducing congestion and air pollution on the roads and the rush-hour crush on public transport, will make travel activity still undertaken much more pleasant and less stressful.

More speculatively, the quality of life, both of the individual and the family, may be improved as a result of reduction in the time spent in longer-distance commuting (and its costs), and the increased incidence of working from home. In addition, with a higher proportion of people's time spent at home or within their locality, neighbours are less likely to be strangers, thereby enhancing community welfare.

What would the UK look like?

What would our surroundings look like, say thirty years into carbon reduction? These are some possibilities, most of them beneficial:

- More trees throughout town and countryside, as far more are likely to be planted not only for their aesthetic role but also their contribution to carbon sequestration.
- An improvement in the farming landscape, with more animals in the countryside owing to the abandonment of high-energy factory farming. More market gardening and local growing of vegetables, fruit and flowers, owing to lower imports.
- More and a better range of local shops which do not need a car to reach them, and many more home-delivery services.
- Less road space given over to cars, both for parking and driving, thereby freeing up space for wider pavements, cycle lanes and cycle parking, seating areas, and so on.
- More people on the streets feeling safer and more getting about on foot and bicycle.
- More children out and about on their own or with friends as parental fears about 'stranger danger' and the risk of injury in road crashes are greatly reduced.
- A different appearance to many buildings owing to the application of a layer of rendering (and insulation) to the outside of the old walls of buildings in order to insulate them cost-effectively, and some roofs of houses and commercial buildings covered with solar PV or panels for solar water heating.
- More energy-efficient groupings of compact housing, with fewer spread-out, low-density detached houses.

Conclusions

The key points of this chapter can be summarized as:

- The provision of an easy-to-use tool which enables individuals to calculate their own carbon emissions. This is the essential starting point to moving towards a low-carbon society.
- The carbon dioxide emissions tool can be used to compare the levels of emissions entailed in travelling by different modes: air, rail, bus and car – or cycling and walking.
- Ensuring global emissions remain under 450 ppm would require carbon

rations in 2030 to be just 30 per cent of the current average. This rate of change, although fast, is not vastly different from the scale of change experienced in personal transport use over the last thirty years (although that was an increase rather than a decrease).

- Introducing carbon rationing would require far better 'carbon information' systems for consumers at the point of purchase, which would then increase the attractions of more efficient, energy-using equipment.

- There are many existing sources of advice which can help individuals reduce their energy use and carbon dioxide emissions.

- As the carbon ration will be phased in over a number of years, this will allow for a gradual process of adaptation – unlike the precipitate shock measure of a fixed wartime ration.

You may still find the idea of carbon rationing unconvincing, perhaps unrealistic. Undoubtedly, it seems far removed from the way society works today. Nevertheless, failure to adopt targets for carbon reduction on a global and national scale and failure to adopt rationing can have only two outcomes. The first is that those who do not yet use their share, mainly people living in developing countries, would have to be prevented from doing so. A simple calculation shows that if each person in the developing world had the same energy-intensive lifestyles as those enjoyed by people in the developed world – and that is the direction being taken – global emissions would now be three times their current level. This would result in an impact on the climate of unimaginable proportions. The second and only other possible outcome is that, together with future generations, the world must witness and bear the costs of devastating and escalating damage from this source into the foreseeable future. Of course, neither of these alternatives can be seriously contemplated. We have to choose a better future.

9. Climate by Chance or Climate by Choice
Conclusions

We are on course to damage the planet irreparably for future generations. Time is rapidly running out for us to make fundamental changes in our lifestyles. However, the message of this book is a hopeful one. Solutions can be found, on a global and local level, to prevent the worst effects of unchecked climate change, and these can be implemented by fair and effective means. There is still time to act – but no time to waste.

What should scare you most

This book has put forward several propositions that may be difficult to accept, including:

- To maintain a safe environment, it will be necessary for there to be an equitable global agreement between all the countries of the world to reduce carbon dioxide emissions.
- A system of carbon rationing is the only way of ensuring that everyone takes responsibility for the consequences of their own emissions.
- The UK's overall contribution to a safe climate will require an annual personal carbon ration which reduces year on year so that by 2050 it will be 80 per cent less than today's average emissions.
- Considerable savings will have to be made by reducing our energy-based activities – it is wishful thinking to believe that the contribution from renewable energy and energy efficiency will be sufficient.
- Air travel is the most polluting form of transport and will have to be drastically reduced.
- The central role of economic growth as the primary measure of progress in our society needs to be set aside in favour of objectives focused on improving our quality of life – without prejudicing the future quality of life on the planet.
- We now know that we are at a defining moment in history: the longer we prevaricate, the more difficult will be the task of solving the problem of climate change, and the greater the regret that we did not act more resolutely and sooner.

While all of these statements challenge our way of life and call for changes that we may not currently welcome, they should be seen as encouraging, not frightening. They constitute the answer – a realistic way out of the predicament that has resulted from our continuing attempts to avoid confronting the realities and problems of climate change.

What is truly frightening is the prospect of insufficient action being taken. The rise in temperature by 2100 could be up to 10°C on land. If this happens in southern England, it will turn it into another Sahara Desert, and there will be huge increases in the violence and frequency of storms, a destructive rise in sea levels and a catastrophic loss of species. This is not simply the nightmare scenario portrayed by fevered environmentalists but the rational analysis offered by climate scientists worldwide. One of Britain's most prominent scientists, Sir John Houghton, has called climate change 'a weapon of mass destruction'. It is already killing far more people than is terrorism but we lack anything like the same political resolve to tackle it. It is climate change which threatens to steadily devastate our quality of life way into the future. The longer we delay, the more severe the penalty – economic chaos guaranteed, social and environmental problems intensified, and the costs of finding workable solutions rising inexorably. Immediate action is needed.

Why rationing is the right answer

There can be no solution to climate change without international agreement to reduce emissions. Contraction and convergence is undoubtedly the right framework for this agreement. It offers equity and certainty of result. Rather than imposing the burden of climate change on the populations of developing countries, and denying them opportunities for a better quality of life, it offers a clear path for development using low-carbon lifestyles and technologies. The countries that are currently polluting most will have to make the largest adjustments but within a framework in which the populations of all countries contribute to the solution. It is vital that C&C is used as the basis for post-Kyoto global agreements – no other negotiating framework offers both 'equity and survival'.

Within the UK, for the half of the energy directly used by individuals, carbon rationing per capita is proposed. It shares the logic of C&C in guaranteeing carbon savings and achieving them in a fair way. There is no alternative credible strategy that ensures the necessary reductions. It is the only politically realistic and morally defensible one that has any real prospect of commanding widespread support. It would therefore be a dereliction of our responsibility both to ourselves and to future generations not to embrace it as a matter of urgency. While inevitably there will be difficulties involved in its adoption and implementation, rationing represents the most equitable approach in this critical area of public policy that affects everyone without exception. It introduces the 'conserver gains principle' (by which those whose lifestyles make a low impact on the environment are rewarded) rather than the 'polluter pays principle' (by which the conscience of those whose lifestyles make a high impact is assuaged by having paid for it).

International negotiations inevitably take some time. But there is no need to wait for a post-Kyoto agreement covering countries in both the developing and developed world before implementing carbon rationing in the UK. The government already has targets for carbon dioxide reduction by 2010 and 2050: rationing could be introduced now to ensure that these targets are met. There is already a worrying policy gap, for it looks unlikely that the current 20 per cent reduction target by 2010 will be met. The introduction of rationing would ensure that we achieve that target and others set in the future to ensure that the energy intensivity of our lifestyles does not prejudice the ecological integrity of the planet. Informing and educating the public, so that they can accept and understand what is involved, should begin at once. The sooner such measures are introduced the better, and the easier will be the transition to the low-carbon future that must be universally embraced.

Actions for individuals

Climate change is too important a problem to be left to the experts to deal with, for it cannot be solved by them alone. It requires the active engagement of all sectors of society – government, the business community and individuals. The UK's carbon dioxide emissions cannot be reduced by the amount necessary without the consent and co-operation of all of us. People need to change their lifestyles but, possibly more importantly, they also need to contribute towards changing society so that a sufficiently low-carbon future is assured. Politicians are not currently giving a sufficient lead on climate change. It is time for a 'bottom up' movement of people to persuade them to take climate change seriously and to act accordingly – and not to hide behind a shield of assumed ignorance or indifference.

The first step that you the reader can take is to complete the emissions self-audit in Chapter 8. When you then compare your total (multiplied by two to account for your share of non-domestic emissions) with our target for world annual per capita emissions of carbon dioxide of 3 tCO_2 by 2030 or 2.1 tCO_2 by 2050, you should be shocked by the contrast.

To focus your attention more sharply on what is required, consider the implications of failure to do so. As we pointed out in Chapter 8, there can only be two outcomes, both of which are totally indefensible. Either the populations of the developing world must be prevented from using their share of the emissions – remember that the planet has only a fixed capacity to absorb greenhouse gas emissions – *or* we will be wittingly contributing to making it largely uninhabitable. Coming to terms with the result of this simple calculation means that we must take the issue of climate change far more seriously in our daily decision-making. There is no escape clause – no 'Third Way'!

What you need to do now

- Set yourself an annual target for reducing your carbon dioxide emissions, using the ration table in Chapter 8. It will be easier to make gradual, sustained changes rather than dramatic reductions that you and your family may find hard to maintain.

- Reduce your carbon dioxide emissions according to your targets. Audit your emissions at least annually to keep track of your progress.
- Take advantage of the many (often free) sources of information about how to reduce household energy use and dependence on motorized travel.
- Keep educating yourself and your friends and family about climate change. Use the references listed in this book and the suggestions under Further Reading to find out more.
- Join or support an environmental group working on this issue and encourage your local community organization / Council / religious group to do more.
- If you are persuaded that the concept of contraction and convergence combined with carbon rationing is the only assured and fair way of preventing awesome damage from climate change, then:

    1. Write to your MP requesting its adoption.
    2. Ask what commitment on the subject he or she will make if elected.
    3. Make your views as widely known as possible.
    4. Get involved in debates on the actions that need to be taken to prevent climate change.

Actions for government

The primary responsibility of governments is to provide security for their people. As the Prime Minister has stated: 'There will be no genuine security if the planet is ravaged by climate change.' The UK government has led the world in setting a target for reducing carbon dioxide emissions by 60 per cent by 2050, and the UK is one of the few countries that is likely to meet its commitment to the modest Kyoto reduction target. Nevertheless, actions to date have not matched up to the rhetoric. Politicians have failed to agree on any realistic programme to avert the global crisis of climate change and, as yet, the present government has held short of endorsing C&C as the framework for delivering the essential reductions of greenhouse gases in an equitable way.

Independent professional commentators doubt that the UK government will meet its domestic carbon dioxide reduction target of 20 per cent by 2010. In common with others worldwide, it has not warned its public of the changes which will be required to prevent dangerous climate change. Instead, it is allowing us to engage in wishful thinking about technological fixes that will enable life to go on more or less as we know it. The government is not doing enough to ensure climate security either in its current actions, or in preparing its citizens for a very different future. On the contrary, and remarkably, it is also making major decisions, for instance on catering for a massive increase in air and rail travel which will make much more difficult the task of achieving the essential targets on reducing emissions.

Government thinking needs to change at a fundamental level. Policy has been formulated on daydreams that economic growth can be sufficiently 'decoupled' from current levels of carbon dioxide emissions so that it can be maintained in perpetuity. In practice, the ecological imperative and political implications of global warming require governments around the world to abandon their consensual belief that greater efficiency and productivity in demand-led economies is the only way of improving the quality of life of their populations.

This poses a major challenge to two conventional assumptions of our time: first, that demand should be met if at all possible. After all, it has always been thought that rising material standards are a legitimate aspiration. And second, that economic growth is essential to prosperity, sustainable within the currently available resources, and the environmental problems that it causes can be resolved adequately. However difficult it is to rethink these assumptions, it will be far easier than dealing with the terrible human and environmental costs of the impacts resulting from further climate change.

What the government should do

- Recognize that the fall in UK carbon dioxide emissions over recent years is largely an historical accident and admit that the UK, while in the vanguard of international commitment on this, has not yet made the necessary fundamental changes towards a low-carbon economy.
- Include aircraft emissions in national accounts of greenhouse gas emissions, and take the lead in moves to do the same internationally.
- Audit its own policies to identify those which are encouraging higher energy use and higher levels of emissions – and then make the necessary policy changes.
- Adopt contraction and convergence as its international negotiating position, and encourage fellow EU countries and other governments that do not already acknowledge the significance of this framework solution to do the same.
- Set 450 ppm as an absolute limit for carbon dioxide concentrations in the atmosphere.
- Start urgent negotiations with all the political parties in the UK in order to achieve a consensus on how to inform the public about the uncomfortable truths of climate change they need to hear and in order to gain support both for C&C and carbon rationing.
- Begin a widespread programme of public education on the links between

carbon dioxide emissions and how our current lifestyles lead to them, providing information and raising awareness as a prelude to the adoption of carbon rationing as the government's strategy for the future.

- Introduce carbon rationing as soon as possible to help meet current targets for carbon reduction.

Our moral choice

Climate change is undeniably the most critical issue that humankind has ever had to face. It is difficult to comprehend the scale of destruction we could bring upon ourselves and future generations. Yet we are heading towards this future with eyes wide shut. Moral purpose must inform our choices: it is not a question of whether or not we need to behave in an ecologically responsible fashion. It is morally indefensible to make decisions that prejudice the imperative of living within the planet's means. We can no longer proceed as if we have an inalienable right to cause damage if there is no other way of achieving our desires.

Responding to climate change is a personal as well as a collective esponsibility. The 'buck' cannot continue to be passed around between individuals, industry, commerce and government. We must act together. Only an equitable approach can provide the right framework to meet the challenge and to have the prospect of success.

We must accept the need for significant modifications to our lifestyles. Technology cannot solve all our problems. We must not allow ourselves to be satisfied by totally inadequate measures, albeit in the right direction. Attempting to anaesthetize our consciences by travelling to the airport by public transport (in order to fly to a distant destination), while genuinely supporting the view that our children should inherit a healthy planet, must be seen as a fundamental contradiction between beliefs and actions. Such 'dissonance' is understandable but unacceptable. We cannot go on indefinitely pretending that our decisions are not making matters worse or making them worse to such a marginal degree that it doesn't really matter. Carbon dioxide emissions that we make now will remain in the atmosphere, affecting the climate for hundreds of years. In this context, well-meaning, symbolic gestures are pointless. What counts is effective action. This book has provided information and ideas on what needs to be and can be done.

Our present and future decisions about the use of fossil-fuels will have a major impact on the quality of life of people in the next few decades and the generations succeeding us. We have a moral responsibility to act with this inescapable truth in mind. Future generations will justifiably sit in judgement on what we chose to do in the early part of this century in full knowledge – as accessories before the fact – of the devastating consequences of continuing with our energy-profligate lifestyles. The accumulation of evidence on climate change and its damaging impacts make it progressively unacceptable that in the years ahead we attempt to plead ignorance with the excuse 'we did not know' – with all its haunting wartime images of the outcome of looking the other way.

Further Reading

Climate change and carbon dioxide emissions

Climateprediction.net: www.climateprediction.net
This site contains a lot of information about climate change and allows you take part in a climate prediction experiment which has been developed to allow a state-of-the-art climate prediction model to be run on home/school/work computers. By getting data from thousands of climate models, the organizers hope to generate the world's largest climate prediction experiment.

Guardian website: www.guardian.co.uk/climatechange
A very good archive of news stories on climate change as well as explanations of the basic science.

Houghton, J., *Global warming: the complete briefing*, second edition, Cambridge University Press, Cambridge, 1997
Detailed explanation and discussion of climate change by one of the world's leading experts.

New Scientist website: www.newscientist.com/hottopics/climate
A very good source of information about the scientific issues concerning climate change, ranging from basic information to the latest research updates. The 'frequently asked questions' section is an excellent place to start.

Energy use

Ramage, J., *Energy: a guidebook*, Oxford University Press, Oxford, 1997
This book gives a very good introduction to energy use, particularly the science and technology aspects.

Contraction and convergence

Athanasiou, T., and Baer, P., *Dead heat: global justice and global warming*, Seven Stories Press, New York, 2002
An American book which covers some of the same material as this one, but from a US perspective. It strongly promotes C&C and discusses in detail the implications of its introduction for international justice.

Global Commons Institute website: www.gci.org.uk
Contains a great deal of material about C&C as well as a model that you can download in order to run your own scenarios.

Meyer, A., *Contraction and convergence: the global solution to climate change*, Green Books, Totnes, 2000
A description of C&C by its originator and how it has gradually become accepted as a leading approach for future global negotiations. Includes good colour graphs showing the implications of C&C for different countries.

Personal carbon dioxide emissions

Recommended reading for reducing your personal carbon dioxide emissions is given in Chapter 8.

Domestic tradable quotas (DTQs) website: www.dtqs.org
Contains details of the work of David Fleming on DTQs since 1996.

References

References are arranged alphabetically by 'author', whether an individual or an organization. Organizations referred to in the main text can usually be located in the references, even if the work is listed under the name of an individual. In those few cases where it might otherwise be impossible to trace a source – for instance, where a quotation or study is attributed to an individual or organization whose name does not appear in the reference – the first few words of the sentence containing the quotation or study are printed in brackets after the reference. See 'Abbreviations' at the front of the book for any acronyms used.

Sources used in more than one chapter

Associated Press, 'France heat-wave death toll set at 14,802', *USA Today*, 25 September 2003.
Blair, T., 'Prime minister's speech on sustainable development', 2003. Published on the internet: www.number10.gov.uk/output/Page3073.asp
DEFRA/DTI, *Energy White Paper: Our energy future – creating a low carbon economy*, Stationery Office, London, 2003 (including 'Recent government statements . . .', ch. 6)
DETR, *Climate change: the UK programme*, Stationery Office, London, 2000
DfT, *Transport Statistics Great Britain 2002*, Stationery Office, London, 2001 – and the same statistical series for earlier years
Transport trends: 2002 edition, Stationery Office, London, 2003
DTI, *Digest of UK energy statistics 2003*, Stationery Office, London, 2003
DTLR, *Focus on personal travel*, Stationery Office, London, 2001 (includes the National Travel Survey)
GCI website: www.gci.org.uk
Hillman, M., 'Why climate change must top the agenda' and 'Carbon budget watchers' in M. Hillman (ed.), 'Special section: climate change', *Town and Country Planning*, October 1998

IPCC, *Special report on aviation and the upper atmosphere*, edited by J. E. Penner, D. H. Lister, D. J. Griggs, D. J. Dokken, and M. McFarland, Cambridge University Press, Cambridge, 1999

Nakicenovic, N., and Swart, R., *Emissions scenarios: special report of the Intergovernmental Panel on Climate Change*, Cambridge University Press, Cambridge, 2000

Olivier, D., *Building in ignorance*, Association for the Conservation of Energy and Energy Efficiency Advice Services for Oxfordshire, 2001

ONS, *Social trends*, various years, Stationery Office, London

PIU, *The energy review*, PIU, Cabinet Office, London, 2002

Ramage, J., *Energy: a guidebook*, Oxford University Press, Oxford, 1997

RCEP, *Energy: the changing climate*, twenty-second report, Stationery Office, London, 2000

The environmental effects of civil aircraft in flight, RCEP, London, 2002

Shorrock, L., et al., *Domestic energy fact file*, various years, Building Research Establishment, Watford

Transport 2000 website: www.transport2000.org.uk

1. Eyes Wide Shut
Introduction

Department of the Environment, *This common inheritance: Britain's environmental strategy*, HMSO, London, 1990

2. Beyond the Planet's Limits
Climate Change: Why, How and What Next?

BBC News, 'North Pole ice "turns to water"', 20 August 2000. Published on the internet: news.bbc.co.uk/1/hi/world/americas/888235.stm.

Campbell-Lendrum, D. H., Corvalan, C. F., and Prus-Ustun, A., 'How much disease could climate change cause?' in A. J. McMichael, D. H. Campbell-Lendrum, C. F. Corvalan, K. L. Ebi, A. K. Githeko, J. D. Scheraga and A. Woodward (eds.), *Climate change and human health: risks and responses*, WHO, Geneva, 2003

Clarke, T., 'Holistic model hints next century could get even hotter than we thought', Nature Science Update, 23 May 2003. Published on the internet: www.nature.com/nsu/030519/030519-9.html ('A new climate modelling approach, developed at the UK's Hadley Centre . . .')

Conisbee, M., and Simms, A., *Environmental refugees: the case for recognition*, New Economics Foundation, London, 2003

Department of Health, *Health effects of climate change in the UK*, Department of Health, London, 2001

Dillon, W., 'Gas (methane) hydrates: a new frontier. US Geological Survey. Marine and coastal geology programme', 1992. Published on the internet: marine.usgs.gov/factsheets/gas-hydrates/title.html

Environment Agency website: www.environment-agency.gov.uk – (search under 'Flood and coastal defence R&D programme' and 'Thames Barrier closures')

Fisheries Research Service, *Climate change, Scottish waters and fishing grounds*, FRS Marine Laboratory, Aberdeen, 2003

Griggs, K., 'New Zealand's belching animals', *BBC News*, 7 May 2002. Published on the internet: news.bbc.co.uk/2/hi/science/nature/1972621.stm

Hulme, M., 'There is no longer such a thing as a purely natural weather event', *Guardian*, 15 March 2000

Hulme, M., Turnpenny, J., and Jenkins, G., *Climate change scenarios for the United Kingdom: the UKCIP02 briefing report*, Tyndall Centre for Climate Change Research, University of East Anglia, Norwich, 2002

IPCC, *Climate change 2001: The synthesis report: a contribution of Working Groups I, II and III to the Third Assessment Report of the Intergovernmental Panel on Climate Change*, edited by R. T. Watson and the Core Writing Team, Cambridge University Press, Cambridge, 2001

Kasting, J. F., 'The carbon cycle, climate and the long-term effects of fossil fuel burning', 2001. Published on the internet: www.gcrio.org/CONSEQUENCES/vol4no1/carbcycle.html

Keeling, C. D., and Whorf, T. P., 'Atmospheric CO_2 records from sites in the SIO air sampling network' in *Trends: a compendium of data on global change*, Carbon Dioxide Information Analysis Center, Oak Ridge National Laboratory, US Department of Energy, Oak Ridge, Tennessee, 2002

Larsen, J., *Record heat wave in Europe takes 35,000 lives*, Earth Policy Institute Eco-Economy Update, Washington, October 2003

Marland, G., Boden, T., and Andres, R. J., 'Global, regional, and national CO_2 emissions' in *Trends: a compendium of data on global change*, Carbon Dioxide Information Analysis Center, Oak Ridge National Laboratory, US Department of Energy, Oak Ridge, Tennessee, 2003

Meacher, M., 'The end of the world is nigh – it's official', *Guardian*, 14 February 2003

Meek, J., 'Wildflowers study gives clear evidence of global warming', *Guardian*, 31 May 2002 ('The records of an Oxfordshire naturalist . . .')

Met Office website: www.metoffice.gov.uk

Munich Re, *Topics 2000: Natural catastrophes – the current position*, Munich Re-insurance Company, Munich, 2000

New Scientist, 'Six months and counting', *New Scientist*, vol. 171, issue 2307, 8 September 2001

'Heat-seeking fish holiday in Cornwall', *New Scientist*, vol. 174, issue 2343, 18 May 2002

Norris, S., Rosentrater, R., and Eid, P. M., *Polar bears at risk*, WWF-World Wide Fund for Nature, Gland, Switzerland, 2002

Press Association, 'Global warming could trigger mass extinction', *Guardian*, 19 June 2003 ('A huge wave of extinction . . . researchers at Bristol University have shown . . .')

Reuters, 'Global warming blamed for melting Everest glacier', 7 June 2002. Published on the internet: www.planetark.org/dailynewstory.cfm/newsid/16308/story.htm

Sample, I., 'Not just warmer, it's the hottest for 2,000 years', *Guardian*, 1 September 2003

Schrope, M., 'Global warming, global fever', *New Scientist*, vol. 174, issue 2349, 29 June 2002

Science and Technology Committee, *Scientific advisory system: scientific advice on climate change*, third report, House of Commons, London, 2000 ('Sir John Houghton, previously co-chairman . . .')

UN, *United Nations framework convention on climate change*, UN, New York, 1992

Vidal, J., 'The darling buds of February', *Guardian*, 23 February 2002

3. As If There's No Tomorrow
Energy Use: Past, Present and Future

Banister, D. (ed.), *Transport policy and the environment*, E. & F. N. Spon, London, 1998 ('Research shows that prolonged car use . . .')

Best Foot Forward, *City limits: a resource flow and ecological footprint analysis of Greater London*, Best Foot Forward, Oxford, 2002

Botting, B. (ed.), *Family spending: a report on the 2001–2002 food and expenditure survey*, Stationery Office, London, 2003 ('Government data show a significant fall . . .')

BP, 'BP statistical review of world energy', 2002. Published on the internet: www.bp.com

DEFRA, *Action plan to develop organic food and farming in England*, DEFRA, London, 2002

DTI, *Energy projections for the UK: Energy Paper 68*, Stationery Office, London, 2000

Garnett, T., *Wise moves: exploring the relationship between food, transport and CO_2*, Transport 2000 trust, 2003.

Jones, A., *Eating oil – food supply in a changing climate*, Sustain, London, and Elm Farm Research Centre, Newbury, 2001 ('The issue of ''food miles'' . . .')

Lane, K. (Environmental Change Institute, University of Oxford), personal communication regarding Oxford University study, October 2003

Lawrence, F., and Millar, S. (eds.), 'Food: the way we eat now', *Guardian*, 10 May 2003 ('A recent calculation showed that a basket . . .')

McNeill, J., *Something new under the sun: an environmental history of the twentieth century*, Penguin Books, London, 2000

National Statistics, *Living in Britain: results from the 2000/2001 household survey*, Stationery Office, London, 2003

Norlen, U., and Anderson, K. (eds.), *The indoor temperature in the Swedish housing stock*, Swedish Council for Building Research, Stockholm, 1993

ONS, 'Access to local services', 2002. Published on the internet: www.statistics.gov.uk/cci/nugget.asp?id=64

'Population trends 110', 2002. Published on the internet: www.statistics.gov.uk

Travel trends: a report on the 2001 international passenger survey, Stationery Office, London, 2002

Open University, *Working with our environment: technology for a sustainable future*, Theme 3, food chains, T172, Open University, Milton Keynes, 2001

PIU, 'Energy scenarios to 2020', 2001. Published on the internet: www.piu.gov.uk/2002/energy/workingpapers.shtml

Pye-Smith, C., 'The long haul, race to the top', IIED, 2003. Published on the internet: www.racetothetop.org/case/case4.htm

Sachs, W., Loske, R., and Linz, M., *Greening the North: a post-industrial blueprint for ecology and equity*, Zed Books, London, 1998

SPRU, 'Foresight futures 2001: revised scenarios and user guidance', 2002. Published on the internet: www.dft.gov.uk/research/foresight/

Sustainable Development Commission website: www.sd-commission.gov.uk

UN, 'World population prospects: the 2000 revision and world urbanization prospects: the 2001 revision', Population Division of the Department of Economic and Social Affairs of the UN Secretariat, 19 December 2002. Published on the internet: www.esa.un.org/unpp

Webster, B., 'Freedom to fly?', *The Times*, 12 August 2002

World Energy Council, 'Global energy scenarios to 2050 and beyond', 2003. Published on the internet: www.worldenergy.org/wec-geis/edc/scenario.asp

4. Excuses, Excuses, Excuses
Personal Responses to the Prospect of Climate Change

Cohen, S., *States of denial: knowing about atrocities and suffering*, Polity Press, London, 2000

Energy Saving Trust, 'Climate change: a 21st Century Ark', 2003. Published on the internet: www.est.co.uk

5. Too Little, Too Late?
Government Policy and Practice

Banister, D., Stead, D., Steen, P., Akerman, J., Dreborg, K., Nijkamp, P., and Schleicher-Tappeser, R., *European transport policy and sustainable mobility*, E. & F. N. Spon, London, 2000

Bell, M., Lowe, R., and Roberts, P., *Energy efficiency in housing*, Avebury, Aldershot, 1996

Carbon Trust website: www.carbontrust.org.uk

Crossrail website: www.crossrail.co.uk

DEFRA, *The environment in your pocket*, DEFRA, London, 2002 ('Government figures show that between 1990 and 2000 . . .')

De Montfort University, 'Consumers left out in the cold when it comes to buying energy efficient new homes', 2003. Published on the internet: www.dmu.ac.uk/news/current/energy_efficient.jsp?ComponentID=9887&SourcePageID=4068

DETR, *Fuel poverty: the new HEES*, DETR, London, 1999

DfT, *Transport ten year plan 2000*, Stationery Office, London, 2000

ENDS, 'Milestone for UK trading scheme', 2003. Published on the internet: www.eceee.org/latest_news/2003/news20030522a.lasso

ENDS Environment Daily, 'More ICT could increase resource use', 23 June 2003. Published on the internet: www.eceee.org/latest_news/2003/news20030623b.lasso

European Environment Agency, 'Environmental signals 2000', 2000. Published on the internet: reports.eea.eu.int/

Grayling, T., and Bishop, S., *The sky's the limit: policies for sustainable aviation*, IPPR, London, 2003

Grove-White, R. (Council for the Protection of Rural England), and Hillman, M. (Political and Economic Planning), *Joint memorandum to House of Commons Select Committee on Science and Technology inquiry on energy conservation*, 1975

Hillman, M., *Conservation's contribution to UK self-sufficiency*, British Institutes' Joint Energy Policy Programme, Heinemann Educational Books, London, 1984

'In favour of the compact city' in M. Jenks, E. Burton and K. Williams (eds.), *The compact city: a sustainable urban form?*, E. & F. N. Spon, London, 1996

'The future of air travel and international tourism', *World transport policy and practice*, vol. 3, no. 1, 1997

'The future of public transport: the dangers of viewing policy through rose-tinted spectacles' in J. Whitelegg and G. Haq (eds.), *The Earthscan Reader on world transport policy & practice*, Earthscan, London, 2003

'The relevance of climate change to future policy on walking and cycling' in R. Tolley (ed.), *Sustainable transport*, Woodhead Publishing, Abington, 2003

Jensen, O. M., 'Visualisation turns down energy demand' in Proceedings of the European Council for an Energy Efficient Economy, 2003. Published on the internet: www.eceee.org

Kuhndt, M., et al., *Virtual Dematerialisation – ebusiness and factor X*, report of a Digital Europe project, Wuppertal Institute (Germany), 2003 ('A recent Europe-wide report . . .')

Myers, N., and Tickell, C., 'The no-win madness of catch-22 subsidies', *Financial Times*, 28 July 2003

Nørgård, J. (Technical University of Denmark), personal communication, June 2001

Pratley, N., 'Anatomy of a budget flight', *Guardian*, 20 August 2003

Simms, A., Oram, J., MacGillivray, A., and Drury, J., *Ghost town Britain: the threat from economic globalisation to livelihoods, liberty and local economic freedom*, New Economics Foundation, London, 2002

Smith, A., and Watson, J., *The renewables obligation: can it deliver?*, Tyndall Centre Briefing Note 4, Tyndall Centre for Climate Change Research, University of East Anglia, Norwich, 2002

Sustainable Development Commission, *Achieving a better quality of life: annual report 2003*, Stationery Office, London, 2003

'Redefining prosperity: resource productivity, economic growth and sustainable development', 2003. Published on the internet: www.sd-commission.gov.uk/pubs/rp/index.htm

'UK climate change programme: a policy audit', 2003. Published on the internet: www.sdcommission.gov.uk/pubs/ccp/sdc/index ('In early 2003, the government's Sustainable Development Commission . . .')

Vowles, J., Boardman, B., and Lane, K., 'Suspecting standby? Domestic levels and the potential for household-level reductions in the UK' in Proceedings of the European Council for an Energy Efficient Economy, 2001. Published on the internet: www.eceee.org

6. Wishful Thinking
The Role of Technology

British Wind Energy Association website: www.bwea.com

DETR, *Waste strategy 2000 for England and Wales*, parts 1 and 2, Stationery Office, London, 2000

DTI, 'Hewitt announces biggest ever expansion in renewable energy', DTI press release, 2003. Published on the internet: www.dti.gov.uk

Edwards, R., 'The nightmare scenario', *New Scientist*, vol. 172, issue 2312, 13 October 2001

Energy Efficiency Best Practice Programme, 'Boiler efficiency database', 2003. Published on the internet: www.sedbuk.com

Energy Saving Trust, 'Powershift FAQs', 2003. Published on the internet: www.transport energy.org.uk/action_faqs_power.cfm

Epple, S., *Towards an energy efficiency strategy for households to 2020: supplementary submission to the PIU energy policy review*, Energy Saving Trust, London, 2001

'European Climate Change Programme – long report', 2001. Published on the internet: europa.eu.int/comm/environment/climat/eccpreport.htm

Eyre, N., Fergusson, M., and Mills, R., *Fuelling road transport*, National Society for Clean Air, Energy Saving Trust and Institute for European Environment Policy, 2002

Factor 10 Institute website: www.factor10-institute.org

Fawcett, T., Hurst, A., and Boardman, B., *Carbon UK*, Environmental Change Institute, University of Oxford, Oxford, 2002

Fawcett, T., Lane, K., and Boardman, B., *Lower carbon futures*, ECI research report 23, Environmental Change Institute, University of Oxford, Oxford, 2000

Foley, J., *Tomorrow's low carbon cars: driving innovation and long term investment in low carbon cars*, IPPR, London, 2003

Gough, C., Shackley, S., and Cannell, M., 'Evaluating the options for carbon sequestration', Tyndall Centre for Climate Change Research, Technical Report 2, 2002. Published on the internet: www.tyndall.ac.uk/research/theme2/ final_reports/it1_22.pdf

Guardian, 'Row erupts over Sellafield security', 7 March 2002. Published on the internet: www.guardian.co.uk ('It has been calculated that if an aeroplane strike . . .')

Hillman, M., 'Ethical implications of climate change for personal lifestyles', *Ethical Record*, September 2001

ICCEPT, *Assessment of technological options to address climate change*, ICCEPT, London, 2002

IEA, *Renewables information 2002*, IEA, Paris, 2002

Johnston, D., 'A physically-based energy and carbon dioxide emission model of the UK housing stock', unpublished PhD thesis, Leeds Metropolitan University, 2002

Leach, G. (Stockholm Environment Institute), private communication on trees which would be required in the UK to offset carbon dioxide emissions, April 2002

Market Transformation Programme, 'Heating appliances in the United Kingdom', 2002. Published on the internet: www.mtprog.com/heating/

Milne, G., and Boardman, B., *Making cold homes warmer: the effect of energy efficiency improvements in low income homes*, EAGA Partnership Charitable Trust, 1997

Olivier, D., *Building in ignorance*, Association for the Conservation of Energy and Energy Efficiency Advice Services for Oxfordshire, UK

Palmer, J., *Final report of the UK national consensus conference on radioactive waste*, UK Centre for Economic and Environmental Development, Cambridge, 1999

Royal Society, 'The role of land carbon sinks in mitigating global climate change', policy document 10/01, 2001. Published on the internet: www.royalsoc.ac.uk

Royal Society and Royal Academy of Engineering, *Nuclear energy: the future climate*, Royal Society, London, 1999

Sustainable Development Commission, *Achieving a better quality of life: annual report 2003*, Stationery Office, London, 2003

'Redefining prosperity: resource productivity, economic growth and sustainable development', 2003. Published on the internet: www.sd-commission.gov.uk/pubs/rp/index.htm

Vehicle Certification Agency car fuel consumption website: www.vcacarfueldata.org.uk

von Weizsacker, E. U., Lovins, A., and Lovins, H., *Factor four: doubling wealth, having resource use*, Earthscan, London, 1997

7. The Solution
Fair Shares: The Only Way

Anderson, K., and Starkey, R., *Domestic tradable quotas: a policy instrument for the reduction of greenhouse gas emissions*, interim report to the Tyndall Centre for Climate Change Research, January 2004

Blair, T., Prime Minister's speech at the G8 Summit in Birmingham, 17 May 1998

Burnett, J., *Plenty and want: a social history of food in England from 1815 to the present day*, 3rd edition, Routledge, London, 1989

Carley, M., Christie, I. and Hillman, M., 'Towards the Next Environment White Paper', *Policy Studies*, vol. 12, no. 1, 1991

CfIT, *A comparative study of the environmental effects of rail and short haul air travel*, CfIT, London, 2001

Christie, I., and Jarvis, L., 'How green are our values?' in A. Park et al. (eds.), *British social attitudes: the 18th report*, Sage, London, 2001

DEFRA, *Environmental reporting: guidelines for company reporting on greenhouse gas emissions*, DEFRA, London, 2003.

Evans, A., *Fresh air? Options for the future architecture of international climate change policy*, New Economics Foundation, London, 2002

GCI, 'Compilation of statements by organizations supporting the contraction and convergence policy proposal', 2003. Published on the internet: www.gci.org.uk/consolidation

Global Commons Net, 2003, website: www.topica.com/lists/GCN@igc.topica. com/read

Meyer, A., *Contraction and convergence: the global solution to climate change*, Green Books, Totnes, 2000

National Energy Foundation, *Simple ways to save energy*, National Energy Foundation, 2003

RCEP, *Transport and the environment: development since 1994*, twentieth report, Stationery Office, 1997 ('The RCEP estimated that an emissions charge . . .')

Rosa, L. P., Muylaert, M. S., and de Campos, C. P., *The Brazilian Proposal and its scientific and methodological aspects*, International Virtual Institute on Global Exchange, 1997

UK government sustainable development website: www.sustainable-development.gov.uk

World Commission on Environment and Development, *Our common future* (the Brundtland Report), Oxford University Press, Oxford, 1987

Zweiniger-Bargielowska, I., *Austerity in Britain: rationing, controls and consumption 1939–1955*, Oxford University Press, Oxford, 2000

8: Watching Your Figure
How to Live Within the Carbon Ration

Boardman, B., Fawcett, T., Griffin, H., Hinnells, M., Lane, K., and Palmer, J., *2 MtC: two million tonnes of carbon*, Environmental Change Institute, University of Oxford, Oxford, 1997

Hassel, C., 'Efficient water management in the built environment', personal communication at seminar at Construction Resources, London, 2002

Palmer, J., and Boardman, B., *DELight: domestic efficient lighting*, Environmental Change Institute, University of Oxford, Oxford, 1998

Semlyen, A., *Cutting your car use: save money, be healthy, be green!*, Green Books, Totnes, 2000

Shove, E., *Comfort, cleanliness and convenience: the social organization of normality*, Berg, Oxford, 2003

9. Climate by Chance or Climate by Choice
Conclusions

Ekins, P., Hillman, M., and Hutchison, R., *Wealth beyond measure*, Gaia Books, London, 1992

Houghton, J., 'Global warming is now a weapon of mass destruction', *Guardian*, 28 July 2003

Index